Erin Loechner's *The Opt-Out Family* is  a world where genuine connections tl Practical strategies and heartfelt passion to empower your kids beyond screens, or. toward a tech-balanced haven.

—**Ginny Yurich,** bestselling author, *Until the Streetlights Come On*; host, *The 1,000 Hours Outside Podcast*; founder, 1,000 Hours Outside

Erin Loechner is like a voice in the wilderness calling out to a new generation of parents. In *The Opt-Out Family*, she invites us into a life that swaps screens for streams, digital programs for developmental play, and social media for real-life connections. As an ardent practitioner of these very principles, she beckons us back to the purpose, peace, and pleasure of a tech-free existence.

—**Ainsley Arment,** author, *The Call of the Wild + Free*

In this book, Erin Loechner manages to simultaneously sound the alarm and offer an encouraging, doable path toward a life well-lived. We can do better by our kids. This book proves it.

—**Sarah Mackenzie,** author, *The Read-Aloud Family*

Through compelling research, insightful conversations with experts, and practical tips, Erin shows us how every time we opt out of technology, we opt in to true life for our families.

—**Brooke Shannon,** founder, Wait Until 8th

A must-read guide for parents searching for life-giving solutions to push against the invasion of screen time and social media in the lives of their kids and families. If you are looking for practical ideas to nurture your kid's childhood wonder, creativity, and independence, this is the book you've been waiting for. Erin is a voice for our generation and provides a perspective that will usher peace, fun, and a new pace into our daily lives.

—**Mandy Arioto,** president and CEO, The MomCo by MOPS International

With growing awareness of the adverse consequences of too much screen time on the developing brain, there remains a deep need to course correct with and for our next generation. Compelling, practical, and full of wisdom, *The Opt-Out Family* helps make a health-based approach to technology use a livable—and desirable—reality.

—**Victoria Dunckley,** MD, integrative child psychiatrist;
author, *Reset Your Child's Brain*

There are few parenting decisions as important to our kids' mental health as how to handle screens. *The Opt-Out Family* will not only tell you what the tech companies will not, but it will give you what you need to cut through the clutter of misinformation and screen normalization that perpetuates so much harm to our kids right now and into their futures. Erin has crafted a book that will help you understand and resist the pull of the unreliable magnetic north of toxic screen culture and help you follow your parenting true north of deep, enduring, and connecting family values.

—**Kim John Payne,** MEd, author, *Simplicity Parenting*; *The
Soul of Discipline*; and *Emotionally Resilient Tweens and Teens*

The Opt-Out Family is a powerful and much-needed antidote for families who are sick of screens dictating their lifestyles and jeopardizing their vital family connections. Erin has figured out, and then written eloquently about, how parents can reclaim and rescue their now-fragile family connections. Using the very tools that created what is jeopardizing family connection to recreate an even better connection is pure genius! The work Erin has done to create this book is a gift and a service to humanity.

—**Pam Leo,** author, *Connection Parenting*

I love this book! Packed with smart ideas, *The Opt-Out Family* gives parents not just the tools to reclaim their lives from screens and Big Tech but also the confidence to do it.

—**Michaeleen Doucleff,** PhD, *New York Times*
bestselling author, *Hunt, Gather, Parent*

Most parents these days are alarmed by the pull that social media, electronic games, and video screens have on our children—and on us. But what can we do about it? The situation seems hopeless. But there is hope, and it is detailed in Erin Loechner's excellent new book, *The Opt-Out Family*. Whether you want to fully opt out or make some small changes, this book will guide you well.

—**Lawrence J. Cohen,** PhD, author, *Playful Parenting* and *The Opposite of Worry*

Erin gives us a path for a slower, more connected life as a family. In the world in which we are living, it won't be easy. But it will be worth it.

—**Denaye Barahona,** PhD, founder, Simple Families

Erin gives weary digital parents exactly what they need: an invitation to discover their children's hearts and allow them to come fully alive. Get this right and nothing on a screen can compare.

—**Chris McKenna,** founder, Protect Young Eyes

If you're wondering why it seems so hard to be a parent and so dangerous to be a kid, and why it's so weird that suddenly we are afraid of almost everything our children do, see, and encounter, you're not alone. You have Erin to guide you through the looking glass of modern parenting. Thank goodness!

—**Lenore Skenazy,** president, Let Grow; author, *Free-Range Kids*

Also by Erin Loechner

Chasing Slow
Courage to Journey off the Beaten Path

The Opt-Out Family

How to Give Your Kids What **Technology Can't**

ERIN LOECHNER

ZONDERVAN
BOOKS

ZONDERVAN BOOKS

The Opt-Out Family
Copyright © 2024 by Erin Loechner

Published in Grand Rapids, Michigan, by Zondervan. Zondervan is a registered trademark of The Zondervan Corporation, L.L.C., a wholly owned subsidary of HarperCollins Christian Publishing, Inc.

Requests for information should be addressed to customercare@harpercollins.com.

Zondervan titles may be purchased in bulk for educational, business, fundraising, or sales promotional use. For information, please email SpecialMarkets@Zondervan.com.

Library of Congress Cataloging-in-Publication Data

Names: Loechner, Erin, author.
Title: The opt-out family : how to give your kids what technology can't / Erin Loechner.
Description: Grand Rapids, Michigan : Zondervan Books, [2024] | Includes
    bibliographical references.
Identifiers: LCCN 2024002525 (print) | LCCN 2024002526 (ebook) | ISBN 9780310345695
    (softcover) | ISBN 9780310368588 (audio) | ISBN 9780310345701 (ebook)
Subjects: LCSH: Internet and children. | Families. | Social media—Psychological aspects.
    | BISAC: TECHNOLOGY & ENGINEERING / Social Aspects | RELIGION /
    Christian Living / Parenting
Classification: LCC HQ784.I58 L64 2024 (print) | LCC HQ784.I58 (ebook) | DDC
    649/.5—dc23/eng/20240309
LC record available at https://lccn.loc.gov/2024002525
LC ebook record available at https://lccn.loc.gov/2024002526

Cover design: Erin Loechner
Cover illustrations: iStock / Adobe Stock
Interior design: Kait Lamphere

Printed in the United States of America

24 25 26 27 28 LBC 5 4 3 2 1

*For the brave, bold, unapologetic parents
and for the revolutionary generation they're not afraid to raise*

Contents

A Note to Readers

This book does not contain success stories of small businesses who got TikTok famous after posting their pastry secrets. It does not contain statistics on how many more teens learned to knit or whittle or grow mushrooms after watching YouTube tutorials in their kitchens. It does not contain happy tales of awkward girls who found their first beloved community on Instagram.

Because for every girl who found herself on Instagram, another lost herself.

We are playing a dangerous game with algorithms. It is a game of chance, of numbers, of wild risk. Daily we roll the dice. If we get lucky, we win. If we don't?

Our children lose.

And we lose them.

Q Opt-out.family   x | 🎤

noun, singular
1. An intentional home environment, group, or household unit in which the role of technology is greatly minimized as a result of evidence-based research on the developmental harms attributed to screens, social media, and digital use

Synonyms: *tech-free family, low-tech household, flip-phone users, device-free, screen-free*

See also: *Opt-out-families: noun, plural*

How to use "opt-out families" in a sentence:
1. Though not all opt-out families live in a screen-free home, many parents have chosen to reject the use of personal devices.
2. A growing number of parents today are identifying as opt-out families, a proactive response to technology's overreach in their children's schools, from data mining to privacy breaches, and mounting evidence of its harmful effects on brain development, attention, and cognition.
3. "'We're an opt-out family,' Mary explained. 'You won't find us on separate screens, scrolling in our own worlds while sitting on the same sofa.'"

Related: see *Opt-out-kids*

Introduction

It Started with a Rope Swing

I was pregnant with my third child, sprawled on the front lawn untangling a steeple rope-ladder for tree climbing. My four-year-old son ran circles around our towering oak, inquiring as to the status of the pending project—*Is it ready? How about now? Like a short soon or a long soon?*—while my eight-year-old daughter and my husband sat cross-legged, each hunched over a particularly tangled knot in the rope. It was a hot July afternoon, and all that remained of our heaping bowl of frozen blueberries was ant soup and violet tongues. My energy was waning. I questioned more than once whether we should abandon the project, save it for a day that promised more ease, less sweat, and definitely more frozen fruit. But we pressed on.

I had a vision, after all.

Our home is a small fixer-upper in a quiet neighborhood with an elementary school nestled right in the center. Daily, my family and I watch the pint-size crowd and their parents rush by with Spider-Man backpacks and weighty dreams. What if ropes and rungs could brighten someone's morning commute? Who might pause? What might he or she discover? What could happen next?

Friends and neighbors halting their day to scramble up to the highest heights of our favorite tree, peeking at the world with a renewed perspective. Fresh air, laughing leaves. Built-in adventure, soaring hopes, wide-eyed wonder. Tea parties hosted on every branch. Tangled hair and, of course, scraped knees. Dirt under fingernails. Hundreds of childhoods claimed and reclaimed. Lives *lived*, if only for the afternoon.

Our community tree-climbing apparatus has swayed in the welcoming wind for three years now. We've hoisted up babies in diapers, spun giddy toddlers, timed competitive third-graders racing an impromptu obstacle course. Our steeple climber has heard shrieks of every variety—glee, exhilaration, then disappointment when a mother announces it's time to head home for lunch. The vision came true.

Until it didn't. Earlier this summer, I noticed our once beloved rope climber had been left vacant, untouched, barely noticed. In just three short Julys, the landscape of our neighborhood (and truly, our world) had changed. Today, I watch teens scoot by on their skateboards, scrolling through feeds or firing off text messages, glancing up every now and then to avoid a sidewalk crack. Parents power walk by wearing wireless earbuds, swiping through emails and podcasts as they push double strollers and juggle lattes. The elementary crowd steers bikes or scooters with one hand, balancing devices in the other as they film stunts for YouTube before rocketing off to some destination.

A few weeks ago, I peered through the window with delight to see a neighborhood child approach the climber with purpose. I witnessed her grasp the ropes, expecting her to heave higher to the next beam, to climb upward and upward until she reached the tip-top of the ladder swing, the canopy of leaves just brushing her eyelashes. But she didn't. She, instead, plunked herself onto the second-to-bottom rung, her feet dangling only a few inches from the weedy grass below. She fished for something in her back pocket, and I realized she'd brought her iPhone along. For the next twenty minutes, I watched her pose, snapping selfie after selfie, finding every single one of them lacking. Over and over, she tried another angle while the tree branches danced high above and chirping creatures scampered from view. This time with pursed lips? Gazing away from the camera, pensive? Touching her hair, bored? Hanging upside down, giggling? Again and again and again. An entire world above her, abounding in beauty, all beyond her reach.

She was eight years old.

The Work of an Algorithm

This is the world our children live in, where playground swings become selfie props. It is not an accident. It is the work of an algorithm. It is the work of a machine and a mission, a grand strategy dreamed up by people in boardrooms who make a living by stealing a life.

Every time you unlock your smartphone, your next few minutes, next few decisions, next few conversations are filtered, shaped, and informed by an algorithm. Your coworker catches you up on his *Yellowstone* binge last weekend, and by the time lunch break hits, your default browser is displaying ads for national park totes and a cheap flight to Wyoming. Heading to a baby shower, you pick up newborn diapers at the grocery store and Target starts sending you coupons for Tucks and Frida Baby—"Congrats! Here's your postpartum survival guide!" Your niece asks Siri the dates of Arcade Fire's next concerts and your Amazon cart recommends the latest summer festival looks. Your son's soccer teammate texts pizza orders on a group thread and his next YouTube recommendation is everything he never knew about #Pizzagate.

Faithfully, we feast on an Apple a day. But the diagnosis is clear: a steady diet of algorithms, smart devices, and tech addiction is changing us to the core. And when we grant our kids an appetizing bite of shiny newness? The consequences are dire.

Developmental delays. Confusion. Loneliness. Stress. Anxiety. Manipulation. Inactivity. Depression. And much worse. According to Tim Kendall, former president of Pinterest and director of monetization at Facebook, "It's plain as day to me—these services are killing people. And causing people to kill themselves" (Allen 2020).

He's right. The dangers of our modern-day digital lives have been well documented, and I'll point to many studies and statistics throughout this book. But the question remains: if smartphones—and their dangerous algorithms—are so terrible for us, why do we engage? Why do we choose to hand today's top addiction to our kids? And why are we, as full-grown adults, sitting on the bathroom floor watching (filming?) TikTok content while our toddlers play alone one room over?

The answer isn't what you think. Of all the expert strategies we can wield against our tech addictions—*Leave your phone in a drawer! Turn off all notifications! Try a digital detox!*—we're addressing only half of the issue. We're silencing our devices. But what silences the desire to turn them on in the first place?

A New Path to Consider

My aim in this book is to offer a new path to consider as you think about your family's approach to digital living. As you'll soon find out, this revolutionary journey doesn't call for parental-control apps, arbitrary time limits, or complicated reward charts. This road travels into far deeper territory, and the map comes from the most unlikely of sources: the algorithm itself.

What if we stole the secrets of the boardroom and the tech wizards and the platform experts? What if we gathered the internet's greatest tips and tricks to guarantee delight online and used them to grow delight *offline*? What if we unpacked every sneaky strategy technology uses to get us to opt all the way in and we flipped the script? Could we rebuild our lost lives? Would we gain back our time and our attention and our energy? Could we learn to be more engaging than the algorithm?

It's a lofty goal to boldly raise an opt-out kid, to design a family culture that is truly yours—entirely yours!—that ensures you'll know the people under your roof better than an Instagram ad does, to create a life that is so good and so abundant and so *full* that you won't find yourself sneaking away from it to check your feeds. But I believe it's possible to live a life that's both full of love and empty of likes.

I *know* it's possible, because I did it.

Something Better on the Other Side

I'm what some might call an internet trailblazer, having started a blog before blogging became a verb, let alone a career. I've been writing to an online audience since 2001, meaning I, along with others who pioneered a similar path, have been influencing various social spaces since well before today's top influencers were out of diapers. We've seen it all—the rise and fall of every tech start-up from About.me to Zoom. Once social-media platforms entered the scene, we were early adopters for most, and predetermined influencers and suggested users for many.

Over the next decade, I garnered a following of 1.4 million people, who pinned images of my bedroom, liked photos of my weeknight sheet-pan dinners, and tweeted excerpts from blog posts I'd written about all manner of subjects: the joys of jigsaw puzzles, the experience of home birth, the importance of ethically made pants. As a six-figure influencer, I headlined conferences and moderated panels alongside Silicon Valley's earliest digital trailblazers. I hosted workshops at Walt Disney World, modeled in Gap ads, and headlined my own twenty-four-episode, two-year renovation show as one of HGTV.com's first premiere web personalities. And then I walked away from it all—from the money, the "fame," and the constant dopamine hits—to search for something better, and truer, on the other side.

How to Read This Book

I've organized this book with the framework of my story in mind. At the height of my social-media career, secrets of engagement were often whispered in hotel lobbies during conference breaks, shared in Q&As after standing-room-only keynotes, and outlined by PR reps and ad agencies to be delivered in a brand's ambassador brief. I have learned every shortcut to capturing (and keeping) your attention. I know how to offer a perfectly enticing call to action, whether you're opening your wallet or voting in the polls. Online, I can summon up any number of award-winning strategies to delight you, entertain you, anger you, embolden you. I have heard much. I have tried some. I have developed many.

These secrets of engagement are touted so highly because they work. When put into practice consistently, the "rules" can offer growth and establish a deeper community. They can communicate mission and purpose. So what if we put them to the test in your household? What if we used them to build *your* dreams, rather than the dreams of a few influencers or big brands or tech bros with three rounds of venture-capital funding?

Would the rules hold up? Could they engage? Delight?
Deliver?

We're going to find out. Each of the following chapters offers a peek at the playbook of some of tech's sneakiest engagement strategies to keep you on your devices. In part 1, we'll identify the problems and pitfalls that come from following the predetermined path of a collective algorithm, rather than considering the individual choices of a family unit. We'll walk through a vision for something better and a strategy for how to get there—stolen from none other than Zuckerberg himself. In part 2, we'll highlight some key engagement tricks online, why these tactics work, and how to utilize these same strategies for your household, with your family, in your very real and meaningful life.* Alongside compelling research and startling insights, you'll find both practical encouragement and creative ideas to establish true and lasting influence within your own four walls—and far beyond.

Throughout this book, you'll find a few resources to guide and direct your family's path forward. You'll notice a LOG OFF label for offline ideas to engage your family, an OPT OUT callout for swapping technology with alternative methods, and a DM section for advice, wisdom, and recommendations from fellow low-tech parents. Periodically, you'll find a +FOLLOW graphic spotlighting nonprofit organizations, advocacy groups, and fellow leaders fighting against technology's influence in the lives of your children.

In addition, you'll notice a SELFIE callout with questions for reflection or prompts to consider as you navigate your family's changing relationship to technology. When you reach these prompts, pause. Five minutes is all it takes to complete each as you consider a path forward for your unique, dynamic family.

Last, this book's companion website is updated frequently to deliver books, resources, articles, interviews, and free tools that offer guidance and inspiration for your journey. You can find it at *optoutfamily.com/more*.

Remember: raising an opt-out kid will look different for everyone. Whether your goal is to keep your toddler entertained without the use of YouTube, to limit your fourth-grader's incessant requests for a smartphone, to shelter your preteen from cyberbullying, to ensure your sophomore doesn't spend her entire paycheck on TikTok's skincare recommendations, or to avoid your own temptation to "just respond to this email real quick," this book will facilitate a larger conversation about why it's not enough to let

---

* When I use the terms *household* or *family*, my hope is that they can serve both broadly and creatively to define the people you walk this planet alongside of, in whatever capacity that may be.

society's preoccupation with technology inform our own families' digital use. After all, the question isn't, "How much tech should we give our kids?" The question, instead, should be, "How many kids will we give over to tech?"

My sincerest hope is that, upon closing this book, you'll have a fully formed vision for what raising an opt-out kid might look like for you. And along the way, step by step, you'll be experiencing an engaged family—*your* engaged family—that can see our digital world as an unwelcome interruption from a vibrant life, rather than the other way around.

Part 1

Chapter 1

The Algorithm

What the Numbers Say and Why It Matters

"Does not he see my ways and
number all my steps?"
—Job 31:4

Tell me how you measure me and
I'll tell you how I will behave.
—Eliyahu M. Goldratt

❌ TECH'S PLAYBOOK

People are numbers.

✔ OUR PLAYBOOK

People are human.

"Can you stay here for one hundred seconds?" my seven-year-old son asks before I sneak out to our detached studio to write this chapter. His feet dangle from the kitchen stool as he sits in front of a plate of sliced apples. His eleven-year-old sister is working hard on a watercolor sketch nearby. I hear my husband down the hallway reading a counting book to our youngest, not yet three. In my home, and in yours, math is at the heart of life's foundational premise.

Our days march forward unforgivingly, ticking quickly on our nightstand alarm clocks, dripping slowly down the wax of our growing number of birthday candles. We take notice of years gone and hours to come: anniversaries, board meetings, dentist appointments. We track how many steps we've walked. Calories consumed, dollars spent, books read. Much can be said about the human race, but perhaps our nature can best be summed by this: whatever is measurable, we measure.

Muhammad ibn Musa al-Khwarizmi, who lived from 780 to 850 CE and is often cited as the father of algebra (*Britannica* 2023), agrees. "When I consider what people generally want in calculating, I [find] that it always is a number" (O'Connor and Robertson, 1999). Even now, the questions I hear about technology prove this deeply human desire to measure, to figure, to know. How much screen time should our toddler have? When's the right age for a smartphone? When do we stop tracking our teen? When do we start? How old is too old? How much is too much? What's the formula for parenting in today's digital age?

Centuries later, just as al-Khwarizmi wisely suggests, we're still after the magic number. And he should know, after all. The word *algorithm* is derived from the latinization of his name.

The Measure of an Hour

Child-development researchers have tried for decades to define (and redefine) healthy boundaries for screen time, social-media use, and online activities. Take, for example, recent guidelines established by the American Academy of Child and Adolescent Psychiatry:

- "Between 18 and 24 months, screen time should be limited to watching educational programming with a caregiver.
- "For children 2–5, limit non-educational screen time to about 1 hour per weekday and 3 hours on the weekend days.
- "For ages 6 and older, encourage healthy habits and limit activities that include screens" (AACAP 2020).

Under these vague guidelines, my toddler can binge-watch *Pinky Malinky* while her ten-year-old sister attempts TikTok's "Benadryl challenge" (which encourages participants to overdose on Benadryl for the hallucinogenic high) and her brother can, just like an eight-year-old boy in Ohio did, take the minivan to McDonald's for a burger after learning to drive on YouTube (Szathmary 2017).

To be clear, I don't fault AACAP for setting a dismally low standard. It's a daunting task to place personalized restrictions around an entity as widely constructed and unregulated as the internet. But I do find AACAP's recommendations, as well as the advice of many of today's digital experts, to be misguided. They present false comfort to mitigate potential damage using the oversimplified metrics of time spent or gained. What is the measure of an hour, anyway? Which is more engaging: thirty minutes spent facetiming with Grandma in Florida or a thirty-minute family walk in the sunshine? It depends on who's measuring.

When experts attempt to place collective value on an individual's experience, we're left with a handful of arbitrary rules stripped of meaning, purpose, and context. But the larger reason why parental controls and screen-time restrictions are ineffective ways to regulate digital use is this: time is only a small fraction of what our future generations are losing. Is your daughter sheltered from a predator because she's on Instagram for just twenty minutes a day instead of thirty? Is your son shielded from cyberbullying because he's offline by eight p.m.? Is your preteen protected from suggested eating-disorder content because she's on only one social-media account and it's private and you check every one of her direct messages?

Even the American Academy of Pediatrics (AAP) feels the weight parents are carrying. A line from AAP's recent digital-media research symposium states that "children now spend more time with digital media than with *any other single influence*" (Shifrin etal. 2015, 6, emphasis added). But instead of lobbying for tighter restrictions, the organization has taken a perplexing approach by loosening guidelines to reflect a more relaxed screen-time approach. Their recommendations now read more like a *Teen Vogue* article, offering vague parenting tips like, "Address digital etiquette," "Be a good role model," and perhaps most yielding, "Kids will be kids!"

This? From one of our nation's most trusted resources for early childhood development? If the experts are as perplexed as we are, it's no surprise we're all scratching our heads over smarter tech use. Meanwhile, the online landscape is evolving in complexity with each passing moment, and we keep giving our kids a front-row seat to the show. The price of admission? *Everything.*

It Is Always a Number

The dilemma is clear: we're relying on yesterday's solutions to solve tomorrow's problems. When it comes to interacting in a digital space, time is no longer the only currency worth considering—for us or for our children. What about attention? What's to be said of innocence lost? Of neurological patterns altered? Of lies believed?

Even the cleverest mathematicians can't pin down a metric for a life utterly changed. But our friend al-Khwarizmi is right. "It always is a number." And the following ones should give us pause.

Brain and Body Development

- Preschoolers who use screen-based media for more than one hour each day are shown to have significantly less development in core brain regions involved in language and literacy (Hutton, Dudley, and Horowitz-Kraus 2019).
- Three months after starting to use a smartphone, users of all ages experience a significant increase in social conformity. Brain scans show that heavy users have significantly reduced neural activity in their right prefrontal cortex, a condition also seen in ADHD, and linked with serious behavioral abnormalities such as impulsivity and poor attention (Hadar et al. 2017).
- Researchers have found a clear link between time spent looking at devices and a greater risk of developing myopia, or shortsightedness, predicting that by 2050, half the world may need glasses (Morrison 2021).

Mental Health

- 46 percent of parents report encountering YouTube videos that were inappropriate for their child's age (Auxier et al. 2020b).
- Children who have been cyberbullied are three times more likely to contemplate suicide than their peers (Van Geel, Vedder, and Tanilon 2014).
- 83 percent of children do not tell trusted adults about abuse they encounter online (Thorn 2022).

Reality Distortion

- More than half of middle schoolers in the United States cannot distinguish advertising from real news, or fact from fiction, believing that "if it's viral, it must be true" (Shellenbarger 2016).
- 40 percent of thirteen- to seventeen-year-olds report it is "normal for people my age to share nudes with each other" (Thorn 2022).

- The number of plastic surgeons with patients undergoing cosmetic surgery for the sake of looking good on social media has quadrupled to 55 percent. The greatest increase is found to be in patients under the age of thirty, particularly teenagers (Rajanala, Maymone, and Vashi 2022).

Physical Safety

- 66 percent is the increase in the risk of suicide-related outcomes among teen girls who spend more than five hours a day on social media (Twenge, Joiner, and Martin 2017).
- 94 percent of children who have received unsolicited nudes from adults did not report the perpetrator, even though 33 percent had previously predicted that they would do so (Thorn 2022).
- In 55 percent of cases where children reported or blocked aggressors, their perpetrators found them again either by creating a new account on the same platform or by using a different platform (Thorn 2022).

Addiction

- 41 percent of teenagers say they are addicted to their mobile devices, and 36 percent say they sometimes wish they could go back to a time when there was no Meta (Jiang 2018).
- 97 percent of teens say they are online daily, including 46 percent of teens who say they are online "almost constantly" (Ellerbeck 2022).

Relationships

- 56 percent of parents say they spend too much time on their smartphones, while 68 percent say they are sometimes distracted by their phones when spending time with their children (Auxier et al. 2020a).

✏ SELFIE

Highlight or underline the statistics in this section that alarm you most.

- Researchers have linked social-media use with decreased marriage quality, positing that couples who do not use social media are "11 percent happier" in their marriages than couples who do (Buscho 2021).
- Children under age fourteen spend nearly twice as much time with tech devices as they do in conversation with their families (Donnelly 2019).

The Real Emergency

The numbers don't lie: social media and an overreliance on smartphone use are, at best, diminishing our lives and, at worst, destroying us. We see it. We feel it. We *know* it. So why do we keep the enemy at arm's reach—in our pockets, next to our pillows, just beside the fruit basket? And why do we willingly hand our kids over to them?

Our excuses are many. We don't want our kids to be left out. We want to teach our kids how to navigate the internet responsibly while they're under our roof. We want to be able to reach them and know where they are. We want to empower them to creatively document their lives. We want to encourage the use of educational apps and learning opportunities. We want them for emergency use, although one could argue—as Dr. Sherry Turkle does in her book *Reclaiming Conversation*—that "the real emergency may be parents and children not having conversations or sharing a silence between them that gives each the time to bring up a funny story or a troubling thought" (Turkle 2016, 26).

➕ **+FOLLOW**

You're not alone. Dozens are fighting alongside you as we work to release technology's grip on society. You can meet them, hear their stories, and join their causes at *optoutfamily.com/more.*

The good news is this: for every reason we are quick to embark on today's tech-trodden path, a dozen creative alternatives abound. You'll hear of many in this book. Some will be simple to implement, others a bit harder. Moving the countercultural way and swimming upstream is no easy task. But know this: at any point, you can course-correct. The digital pond we all once splashed into has given way to a river of rage, and no matter your kids' ages, it's never too late to throw in the towel, jump ship, or whatever plucky water metaphor you can manage. Remember, tech's tricky playbook is what got us all into this mess. Now? We'll use its secrets to get ourselves—and our kids—out of it. And of this I'm certain: all will be worth it as we fight to reclaim the heartbeat of our families.

The Highest Stakes

In his book *Irresistible: The Rise of Addictive Technology and the Business of Keeping Us Hooked*, Adam Alter writes, "According to Tristan Harris, a 'design ethicist,' the problem isn't that people lack willpower, it's that 'there are a thousand people on the other side of the screen whose job it is to break down the self-regulation you have'" (Alter 2018, 3).

It happens like this: a generically hip man with rolled-up sleeves, rimmed glasses and a certain degree of facial hair will begin a keynote speech on his dream for greater connectivity, productivity, and purpose. He'll outline the inner workings of his recent algorithm, taking great care to highlight its measurable benefits—increased engagement, more time on device, "happier" users—and taking even greater care to gloss over any language or results that call out this grandiose vision for what it is: a global social experiment of the highest stakes. The blind leading the blind, the beards growing more beards.

Once, at Chicago Tech Week, a friend who is a leading search engine optimization strategist tries to explain to me what, exactly, an algorithm does—and why. "An algorithm," he tells me, "is like a recipe. It's a set of rules. A list of tasks that, if you follow them, will point you to a desirable result."

"Desirable for who?" I ask.

"And that, Loechner," he laughs, "is the million-dollar question."

Want to be a better salsa dancer? Here's a cumbia algorithm.

Want to get your kid into an elite college by the age of sixteen? Here's an Ivy League–track algorithm.

Want to distract, addict, manipulate, and capitalize on an entire society? Here are a thousand algorithms.

But for now, let's unpack just one of them: Instagram. What began as a photo-sharing app back in 2010 where simple shutterbugs could document their days has, since, morphed into a dangerous breeding ground of misinformation, sensationalized content, and an entire generation's plummeting mental health. So how did it get there—and so quickly?

The answer lies hidden in the algorithm.

A Nightmare Machine

Imagine you're a new parent to a new baby—a sweet, cooing little son with Gerber Baby eyelashes and a dimple in his chin. You're smitten, and so are your loved ones. You decide to use Instagram as an easy way to update faraway friends and family on your son's latest milestones, from pureed peas to preschool paints and everything in between. A baby album for the digital era, so to speak.

This is what tech writer Geoffrey Fowler did after his son was born. But then, things got weird.

Shortly after Fowler's first post, Instagam started showing Fowler images of blistered babies, hospitalized babies, intubated babies. Photo by photo, his sweet new account was warping into a fun-house mirror of every parent's innermost anxieties. To find out why, Fowler dove deep into algorithm research, arriving at the startling truth that has now been confirmed in news outlets, congressional hearings, and more than twenty-two thousand pages of internal Facebook documents (Brown 2022): his experience is not the exception but the rule. "This is how the software driving Instagram, Facebook, TikTok, YouTube and lots of other apps has been designed to work," he writes. "Their algorithms optimize for eliciting a reaction from us, ignoring the fact that often the shortest path to a click is fear, anger or sadness" (Fowler 2022).

Stronger reactions create stronger bonds. *Addictive* ones.

This algorithm is the same one your teenage daughter uses to find new vegetarian recipes. It's the algorithm your son uses to share video bloopers. It's the one you use every time you post pictures of the first day of school, your weekly Target run, the pesto recipe you mastered. And it's the same one that sends everyone else's photos to you, *but only the ones it predicts will elicit a stronger reaction than the one before.* This is how, over time, your daughter's harmless vegetarian recipes can guide her to startling pro-anorexia content. This is how your son's feed is filled with extreme selfies and daring experiments. And this is how you can hop onto Instagram for a few minutes and find yourself knee deep in anxiety, fear, and overstimulation a full hour later.

The Gatekeeper of Our Experience

It wasn't always this way. In 2017, at an invitation-only tech conference in New York City, Thomas Dimson, former director of engineering at Instagram, offered a behind-the-scenes peek into Instagram's newly released—and highly controversial— experimental algorithm for a select few data scientists, tech writers, and researchers. Dimson explained that this updated algorithm was designed to keep users on the platform longer, to find content and users they'd enjoy more easily and more quickly. To employ this new experiment, Instagram rearranged every user's feed from a chronological timeline of their friends' photos to a machine-learning algorithm of the nation's photos—and beyond. Why?

So the algorithm can show, in Dimson's words, "the moments *we believe* you will care about the most" (Dimson 2022, emphasis added).

Gone were the days when a new parent could snap a photo of their sleeping infant to share with a few friends in real time, then log off to care for said infant. Instead, Instagram stepped in to flood a new parent's feed with more ideas from more people selling more things—breast pump recalls, vaccine stats, lists of "ten things you should never *ever ever* say to a new parent!"—eliciting a predictably emotional response any way they could get it.

And in that single update (Hackett 2023), Instagram shifted from keeper of our photos to gatekeeper of our experience. Or, perhaps, *owner* of our experience. Without your knowledge or consent, Instagram's current terms of use grant Instagram permission to use any content at their discretion. "When you share, post, or upload content that is covered by intellectual property rights (like photos or videos) on or in connection with our Service, you hereby grant to us a non-exclusive, royalty-free, transferable, sub-licensable, worldwide license to host, use, distribute, modify, run, copy, publicly perform or display, translate, and create derivative works of your content" (Meta 2022). Your Smoky Mountain family vacation, your daughter's first dance recital, your son's lacrosse team highlight reel? If they're on Instagram, they're not your memories. They're Instagram's cash flow.

As Dimson outlined Instagram's newest strategies for maximum engagement, I recall a moment in which he paused his talk to offer a rare humility for the magnitude of this work. "People are asking what actually is going to happen when you take this limited experiment and you, like, deploy it at scale. And as the person who was tech lead for this [algorithm], I was like, 'Oh crap, I don't actually know what's going to happen'" (Dimson 2022).

What happened was this: Instagram's revenue skyrocketed from $1.8 billion under the old algorithm to an estimated $47.6 billion (Iqbal 2023). Users kept using. More people joined. Friends reported seeing fewer photos from the people they were following and more photos from people Instagram thought they *should* be following. Influencers scrambled to make it to the top of everyone's feeds, and in brand briefs and marketing meetings, advertisers encouraged influencers to "beat" the algorithm by producing more reels, taking better selfies, virtue signaling on social justice issues, writing more emotionally charged captions, and creating more "authentic" content (for example, crying in your car). "Get outlandish! Be unforgettable! Live your truth!" was the new mantra for anyone chasing followers on the platform.

As you can imagine, the consequences were dire. We, as a society, have historically made decisions—from shades of lipstick to political affiliation—based on the information available to us. But what happens when we don't choose the information? What happens when an autonomous algorithm chooses *for* us, cherrypicked from a bushel of peers scrambling to be the ripest, newest, most unforgettable star?

What happens is this: we assume that the highlighted options (or "moments," as Instagram loves to call them) are the best, or only, ones that exist. We take the shortcut toward truth. Our world gets smaller. So, too, do our lives. As Dimson himself noted in his NYC talk, "What's optimal for the viewer is not really optimal for [Instagram]." And with a revenue growth of more than $45.8 billion in a few short years, it is clear which entity is more highly valued.

They Will Be Your Neighbors

So what are we going to do about it? Now that we know the numbers, now that we see what's at stake, it's impossible to look away. At least, it was for me.

Once upon a time, my livelihood relied on maintaining a social-media presence. As an influencer, I was successfully running a six-figure platform by writing and photographing multimedia campaigns for Target, Martha Stewart, Pinterest, and hundreds of other brands you know, use, and love. I taught digital workshops in a boutique studio of Singapore, modeled a slow fashion campaign in the Taj Mahal, and twirled pasta with Maria Shriver and Hoda on *The Today Show*. But with every algorithm change—from loosened privacy policies to tighter regulations on speech—a niggling feeling crept into my consciousness: How much more will I feed to this machine? And what will it spit out in return?

As mathematician Dr. Hannah Fry writes, "Whenever we use an algorithm—especially a free one—we need to ask ourselves about the hidden incentives. Why is this app giving me all this stuff for free? What is this algorithm really doing? Is this a trade I'm comfortable with? Would I be better off without it?" (Fry 2019, 47).

My answer came slowly, and then all at once.

In 2014, a *New York Times* photographer visited my home to shoot a spotlight story on the future of blogging, influencing, and new media. She chased my toddler around the house, talked Japanese tea rituals with my husband, shared pistachios on the back porch. As she left, we exchanged numbers. With a hug goodbye, she said this: "I'm rooting for you, girl. Tech's crazy, and it's about to get crazier. Be sure to keep the real stuff real good."

What she was saying, I think, is that there will come a point for us all when we must question how much of our real, extraordinary, multidimensional lives we will steamroll to fit it all into the flattened surface of a rectangle that dings. Her cautionary words helped propel me to a place where life comes before likes, a place where it's ever obvious how much of our world we're giving away to these temporarily fulfilling (entirely virtual) apps, and how little we're gaining in return. Time. Attention. Energy. Brain space. *Life.*

To begin my personal tech revolution, I deleted all apps from my smartphone. (It turns out banking in person can be quite satisfying, and the kids get a lollipop out of the deal.) My husband and I both experimented with a flip phone, then no phone, then eventually landed on a family hybrid of all three. But social media was gone for good. One platform at a time, I stopped using the accounts

I'd built, ones that, in total, reached more than one million people. You might wince at that number, and there was a time I might have, too. *Think of the collective impact! The influence for good! The money and opportunities you're losing!*

But I know firsthand of the far greater fruits gained. Take it from me: you can build a digital home on the borrowed real estate of any social platform. But you'll never own it. It will never be yours. The maintaining *(Sorry, just have to take a quick pic!)*, the beautifying *(Which filter should I use?)*, the cropping *(I hate that angle)*, the producing *(Wear your matching pj's!)*, the creating *(Hmm, what to caption this . . .)*—what's it all worth when the storm comes in? What value does a facade of a life bring? What comfort can it offer—to you or anyone else? As Jenny Odell beautifully challenges in *How to Do Nothing: Resisting the Attention Economy*, "Let's not forget that, in a time of increasing climate-related events, those who help you will likely not be your Twitter followers, they will be your neighbors" (Odell 2019).

Something Better

Herein, then, lies our choice. We've been given a cursory glance at what's at stake. We've analyzed the metrics. We've run the numbers. And now we get to move forward.

If an algorithm is little more than a recipe for a result, what's Silicon Valley cooking? A lot. Meta's CEO Mark Zuckerberg kicked off his keynote presentation at Facebook's 2021 Connect conference by detailing his ambitious plans for building out what he calls the metaverse—a 3D virtual space where people from all over the world can meet at any given moment. "In this future, you will be able to teleport instantly as a hologram to be at the office without a commute, at a concert with friends, or in your parents' living room to catch up. . . . Think about how many physical things you have today that could just be holograms in the future. Your TV, your perfect work setup with multiple monitors, your board games and more. . . . You'll move across these experiences on different devices—augmented reality glasses to stay present in the physical world, virtual reality to be fully immersed, and phones and computers to jump in from existing platforms" (Zuckerberg 2021).

Fake thunderstorms. Fake whale songs. Fake wild flowers and lemon drops and tea kettles. Is this truly where the algorithm leads? In his NYC talk on his new machine learning code, Thomas Dimson quipped this: "At the end of the day, what we want people to do is to come to [Meta] to be happy, enjoy their experience, and come back more."

Tell me, is that your goal as well? Is this the future you want for your children? If so, then your phone has just the recipe.

But if not, keep reading. There's something better and truer—for you and your family—just around the bend.

✏️ **SELFIE**

Revisit the statistics you highlighted in this chapter. Which are most concerning to you, and why? In a spare notebook or companion journal, write out your most pressing thoughts on the role of technology in your life and home today.

Chapter 2

Disrupt

Casting the Vision for a Better Tomorrow

We must not fixate on what this new arsenal
of digital technologies allows us to do without
first inquiring what is worth doing.
—Evgeny Morozov

Let no one deceive himself.
—1 Corinthians 3:18

TECH'S PLAYBOOK

 Technology is the future!

OUR PLAYBOOK

Is technology the future we want?

I hadn't meant to leave social media permanently. Facebook, Twitter, Instagram—a decade ago, these were all positioned as the internet's water cooler, a place for people all over the globe to fill up their glasses, make small talk, ask of the weather, the news, the deadlines. We briefly exchanged a few stories, funny jokes, cute cat photos. Social media was just another pause in the day, a break to rejuvenate before we sauntered back to our desks, refreshed and ready to take on another task.

What social media soon became wasn't a water cooler but a fire hose. Algorithms learned our behaviors and budgets, the people we knew and the ones we wanted to know. Ads took hold, followed by suggested accounts and users and endless recommendations for more to see, to do, to buy, to be. Anyone truthful enough will admit to the unknown number of hours they've lost in the depths of social media, where they learned that finishing that chapter or chorus or proposal was far less important than discovering new ways to organize their pantry. Instinctively, we knew we'd be far more productive if we ditched the platforms altogether. But like most users, I waffled. This was my job, I justified. In today's creative marketplace, doesn't everyone need social media? Isn't social media important for all the things we say it is: platform building, audience connection, new opportunities? I didn't know.

I do now.

I quit Facebook more than fifteen years ago. As online talent for HGTV.com, my husband and I were used to receiving random requests, encouraging messages, and even strange solicitations through the platform. But there was one incident, a particularly harrowing one, that opened my eyes to the vulnerability that comes with having a public Facebook account. My profile was cloned and taken over by some unknown person who used it to reach out to my friends and acquaintances, asking for their addresses, requesting they meet me at different places in town. A stranger pretended to be me, downloading photos of my nephew at the zoo, my cousin and me backpacking through Ireland, recaptioning and reuploading them to their own account. I was so young, maybe twenty-five, so unsure of what to do or what dangers might lurk. I deactivated shortly after the profile was reported and removed, forever unsettled. Even today, when I hear of parents posting images of their children to Facebook or any social-media platform, I know the risk. How easy it is to steal an identity, or worse. How sickening it felt to see a photo of me in the hands and at the mercy of someone else—stolen, uploaded, and used by TheRealErinLoechnerme. Me, but not me. Happy, vibrant, smiling widely in front of the Getty.

Twitter was next. I was hacked, shortly before the release of my first book. A gaggle of unruly teens thought of a genius way to make money—take over any account with a blue verification badge, hold it hostage, see what happens next. Twitter support was, of course, unhelpful in regaining access, but it didn't matter. I found that after months away from the platform, I didn't want it anymore. I didn't miss tiptoeing into current events only to drown in hot takes and fast news and all-caps shouting. Twitter felt like a schoolyard without a supervisor, where inevitably a bully interrupts the kickball game, throws the ball over the fence, and everyone, after a brief shouting match, shuffles home to find something else to do.

Two down, one to go. Here, then, was Instagram, the platform I knew would be hardest for me to untether from. I was an early adopter when it was released in 2010, when everything you shared was somewhat cryptic—your shoes or your dog or a latte, all cloaked in a Valencia filter. Photos of work trips and HGTV projects and city lights—it was all there, a whole history, and if I scrolled back far enough, I could see hints of my pregnancy, my first baby, the beginning of a family. I had established a close-knit circle of Instagram friends, or something like it. Could I leave that community behind? Did I want to?

You Are Implicit

But then, a shift. It happened so insidiously that I would be hard pressed to come up with an exact timeline. But I felt it, a turning of the tide. Under the careful watch of Meta, Instagram began to rage, choppy with waves of riots and controversy and division. I might have argued, at the time, that the platform was merely a reflection of what the world was already like in the throes of a global pandemic. But how would I know? Everyone I knew was on lockdown. I couldn't see beyond my front door, past the mailbox, down the street. I trusted social media, the news pundits, the internet to tell me what lay beyond.

We know now that those waves were heightened, that while those were very real and tumultuous times, it was our anger—not our empathy, not our grief, not our shared humanity—that kept our attention rapt, which, of course, is what Meta wanted.

The rules were clear for anyone with any established platform: if you stay in the ocean, you must ride every wave. You must witness every injustice, you must comment on every headline, inform yourself in hours or less. Your words must come out perfectly. If you don't share the right thing at the right time, an apology will be demanded. But if you share nothing, if you remain silent, you are implicit.

Instagram was no longer a welcome place to talk about paint shades or classic novels or your grandmother's banana-bread recipe. We must speak only of here, now. We must bear witness to the collective fury of corrupt presidents, superspreaders, Karens and male Karens. Social media is a welcome place only for those who represent the common good or whoever's voice joins the loudest song.

Instagram, for a time, felt like a place where you were required to employ a speech writer, a fact checker, and a public-relations representative to maintain any shred of dignified presence. And perhaps it should have always been this way. Maybe it would have been better to tread lightly from day one. Perhaps we took for granted what it means to be a mouthpiece—just one of many—in a society that relies on shared language for healthy progress. Perhaps many of us failed to measure the weight of our influence.

But I think what is truer is that many of us failed to measure its cost.

Walking away from social media was not a calculated, strategic decision. I did not announce my last post, did not attempt to siphon my followers to a sales funnel, a newsletter, a blog, or yet another subscription platform. I just stopped logging on. I rejected the idea that I needed to devote another minute to performance, that—in an effort to avoid cancelation—I must maintain the image of someone who is perfectly informed of activism and science and human rights while also rightly raising humans, nonchalant and carefree in all the ways that are socially acceptable (iced coffee + hot mess, you know the drill).

I suppose what I'm saying is, I canceled myself.

Another Way to Exist

After a few weeks, then months, then years of Instagram absence, I knew I wouldn't be back. I loved not having another place to be responsible to (truthfully, I have a hard enough time tidying my pantry), to show up for, to keep track of. I loved living in a perpetual state of questions, of uncertainty, of wonder, not yet having formed a complete viewpoint to accompany a filtered photograph and corresponding hashtags. There is freedom in the unknown, the uncaptured. In the absence of cropping and publishing everyday life, I found that what was left was mine alone.

Today, surprisingly, I find myself more informed, more engaged, more curious than ever. In the midst of conversations over coffee or at the park, I am no longer racing my way to the point of it all. I am meandering, asking, connecting dots that Twitter pundits haven't attempted to connect for me. I am observing, rather than watching. I am learning, rather than scrolling.

I am living a life, rather than broadcasting one.

Months later, I told all of this to a woman on an airplane, and she stared at me, dumbfounded. "Sounds really nice, and all, but how will people find you? Don't you want to keep working? If my daughter isn't online," she says, "she doesn't exist."

"Then she must find another way to exist," I told her.

Goliath Is David

I first learned the term *disruption* in the Los Angeles office of TBWA/Chiat/Day—the ad agency responsible for Apple's multiple-award-winning "1984" ad. You've likely seen the ad or heard of it. The short version is this: In a gray, bleak, dystopian world, a young female wearing a track-and-field outfit sprints away from officers in riot gear who are chasing her. Long lines of homogenous citizens march in unison down a long tunnel of telescreens and monitors. Armed with a giant sledgehammer, the young woman races toward a large screen on which a Big Brother–type leader is giving a speech promoting the unification of thoughts, pure ideology, and information-purification directives. Hurling the hammer, she destroys the screen in a fiery explosion of light and smoke. The citizens, now jolted awake, prepare for the revolution of a new world.

Most of us know what happened next. Positioning itself as the representative underdog arriving on the scene to empower individuals, Apple went on to sell $3.5 million of MacIntoshes just after the ad appeared. But many don't know what happened just before the ad ran. TBWA/Chiat/Day created the ad from an earlier draft of an Apple print campaign that read, "There are monster computers lurking in big business and big government that know everything from what motels you've stayed at to how much money you have in the bank. But at Apple we're trying to balance the scales by giving individuals the kind of computer power once reserved for corporations" (Burnham 1984).

Sound familiar? Computers lurking. Big business, big government. Knowing everything. *Power.* As a child of the '80's, I was an infant when this ad ran. But it didn't matter; the messaging stuck. As I grew from diapers to Digi Pets, I breathed the same oxygen my parents' generation breathed: technology will save us. *Just think: a computer in every house! Free internet in schools! Information at our fingertips!* Like a well-meaning David slaying an evil corporate Goliath, Apple stepped up to fight for the rights of a more powerful, more informed individual. *We're on your side! Together, we can defeat the monster computers lurking!*

Now, four decades later, the new Goliath is David himself. At the time of this writing, more than two thirds of your iPhone apps are sending personal data about you—and your children—to the advertising industry (Davies 2022). "What shocked me about this is that we have a law in America that's supposed to protect the privacy of children—and yet this is happening," says tech columnist Geoffrey Fowler. "But the problem is that this giant industry of app developers, and also Apple and Google who run these app stores and make billions of dollars off of it, have found some really big loopholes in that law. So they're doing it anyway."

The result? By the time your child turns thirteen, your phone will have sold roughly 72 million data points about him or her—where they shop, what bands they listen to, how they style their hair, their favorite sneaker brand, their religion, their eating habits, their address, their phone number, their vacation photos—to any and every company imaginable (Davies 2022).

⟳ OPT OUT

Don't post photos of your kids online. Resist the urge to make your family your feed, no matter how cute those matching pajamas look. Our children needn't be our platforms, we must be *their* platform. Be their foundation, their support, their bedrock as they meander through life without a camera trailing behind them. Protect their childhood. Give them the expansive, uncommon gift of a future where they can walk forth making their own digital footprints, if—and only if—they so choose.

"I want to yell at that liberatory young woman with her sledgehammer: 'Don't do it!' writes historian Rebecca Solnit. "If you think a crowd of people staring at one screen is bad, wait until you have created a world in which billions of people stare at their own screens even while walking, driving, eating in the company of friends—all of them eternally elsewhere" (Solnit 2014).

Who, then, will disrupt the disruptor? The opt-out family.

Something Truer Is Born

One of my first job interviews as a young twentysomething living in Los Angeles was with, ironically, the agency who sold us all on the dystopian dream: TBWA/Chiat/Day advertising agency. I arrived a bundle of nerves: overwhelmed, underqualified, and supremely overdressed. Sitting stick-straight in an Oxford shirt and a pencil skirt, I waited in the common area just past the indoor basketball court, the green space, the open industrial staircases winding every creative through a seemingly new dimension. Everything was modern, new, fresh. And very, very yellow.

As teams of executives and writers skittered by, hyped up on caffeine and ideas, I watched the account manager approach me. Warmly, she held out a mug and a book. "This is for your nerves," she said, handing me green tea. "And *this*," she said, holding up the book, "is for everything else."

I read the title (yellow, of course): *Disruption*. Inside, I was introduced to the author and chairman of TBWA Worldwide, Jean-Marie Dru, who coined the term *disruption:* "Disruption means identifying, questioning, and overturning the conventions that define an existing . . . situation. Such conventions exist, but they are usually hidden. Conventions are the unquestioned assumptions, the common sense wisdom, the current rules of the game that comprise the status quo. . . . The job of Disruption is to uncover these conventions, question their validity, and displace them" (Dru 1996, 62).

I eventually came to understand that this book is a sort of Bible in the tech advertising world, outlining the highly successful strategy behind Apple's biggest marketing campaigns. The three-part process works like this: convention, disruption, vision. Through a series of prompts and exercises, TBWA/Chiat/Day's brainstorming teams dismantle years' worth of assumptions, misconceptions, and traditionally status-quo ideas. And then, something truer is born.

The process has given way to hundreds of daring ideas, creative solutions, and innovative plans. So why not put it to work for us, in our homes and families? Why not disrupt Apple's sleek army of products: iPads, smartphones, and the many apps that they rely on? It's time to face our assumptions about technology's role in our kids' lives. It's time to decide whether the future that has been sold to us is one that we actually want.

1. *Convention.* Our conventional thinking is this: kids need smart technology. Why? We cite many reasons. To foster social connections with their peers. To provide safety in an emergency situation. To grant independence and greater responsibility. To learn healthy tech boundaries. Whether it's for work or play or something in between, we don't question our assumptions, deeming smartphones a necessary tool in the lives of the next generation. But are these reasons true? Let's look at the validity of these assumptions.

2. *Disruption.* The quickest way to question our assumptions is to ask a single question: What if? What if this thinking isn't correct? What if this reason isn't valid? Following, I'll take our conventional reasons as starting points for the disruption process.

Smartphones and Social Connections

▶ **Convention:** Smartphones help children foster social connections with their peers.

▶ **Disruption:** What if smartphones don't help children foster social connections with their peers?

While it's widely believed that today's generation is head over heels for social media, many teens are quietly stepping away from the rat race of constantly snapping, scrolling, texting, and tiktoking. According to a recent study on Gen Z by marketing firm Hill Holliday, half of those surveyed stated they "had quit or were considering quitting at least one social media platform" (Kale 2018). When it comes to kids' desire to maintain relationships on social media, "significant cracks are beginning to show."

"It's just such an interruption," said a fourteen-year-old girl I spoke with who is glad she doesn't have social media. "I'll be trying to talk to my friend about struggles or problems, and, like, her notifications will go off about something she posted. It totally derails the conversation and then we have to talk about, like, what this one person thinks about her selfie."

A seven-year-old boy echoed the sentiment. "My mom puts pictures on Instagram," he says. "It takes a lot of time. She's always apologizing for how long it takes but she still does it. It doesn't look very fun. I'd rather play basketball and not take pictures."

His older brother, seventeen, quit social media for good. "I love it," he says. "I still have friends, they just have to, like, text or call me. But I think it's made my relationships deeper. It's, like, the people that seek me out are real friends. I don't want to keep up with eight hundred people, you know? That's stressful, man."

A youth pastor I spoke with recently said he permanently deactivated the church's youth-group Instagram account. "I realized I was having kids join Instagram just to keep up with when/where we were meeting," he said. "After I let them know I was shutting it down and would text them instead, a bunch of kids came up and thanked me. Turns out they didn't actually want to be there, either. They just were because everybody else was."

These aren't standalone examples, and the science confirms it. Roberta Katz, a senior research scholar at Stanford's Center for Advanced Study in the Behavioral Sciences, recently published the results of a multiyear research project on Gen Z's thoughts, behaviors, and ideas. Citing this generation as the first to adopt an early relationship with powerful digital tools, the head researchers assumed the focus group would reference some of their favorite technologies. "We expected the interviewees to [choose] text, email, chat group, DM, FaceTime, Skype, etc.—but instead nearly every single person said their favorite form of communication was 'in person'" (De Witte 2022).

A Louisiana elementary-school teacher recently witnessed a

striking similarity in her classroom. After she gave the class a writing prompt, four of her students told her they wished phones had never been invented. Wrote one student, "I don't like the phone because my [parents] are on their phone every day. . . . I hate my mom's phone and I wish she never had one" (WKBW 2022).

So why do we still assume social media is the preferred way of fostering connection? Why are we failing to listen to a generation that tells us the opposite: that social media is a distraction, an interruption, a hindrance to the very thing we assume it promotes?

Because of a key barrier to the disruption process: *argumentum ad populum*, which is a logical fallacy that affirms that something is good or true simply because the majority of people *think* it is. And it's the road that leads to a collective status quo on a global scale.

If you were to ask the public whether social media makes them feel more connected, they'd likely respond with an emphatic yes. But if we approach that question critically, what is it they're feeling more connected to? Knowledge, sure. Information. Entertainment. Ads. News.

But people? Each other? Loved ones? That's not what the research shows. After all, there's a reason it's called social media. It's *media*.

As MIT professor Sherry Turkle notes in David Sax's book *The Revenge of Analog*, "Sociable technology always disappoints, because it promises what it cannot deliver. 'It promises friendship but can only deliver performance'" (Sax 2016, 240).

Smartphones and Emergencies

- ▶ **Convention:** Smartphones provide safety in an emergency situation.
- ▶ **Disruption:** What if smartphones don't provide safety in an emergency situation?

TikTok can't dial 911. Twitter won't throw you a life raft. And Instagram won't shield you from a mass shooting. Ken Trump, president of National School Safety and Security Services, cautions that cell-phone communication can actually increase safety risks in the event of a school shooting (Walker 2023). "During a lockdown, students should be listening to the adults in the school who are giving life-saving instructions, working to keep them safe," he says. "Phones can distract from that. Silence can also be key, so you also don't want that phone noise attracting attention." Even among adults, smartphone distraction has been proven to hinder our ability to react swiftly in an emergency. Security-camera footage from a San Francisco light-rail train reveals that a gunman was able to pull out his gun and "openly handle it at length without anyone noticing, before he eventually shot a fellow passenger" (O'Connor 2013).

The truth is, we should be far less concerned about what a smartphone might offer in an emergency situation and far more concerned about what kind of emergency situation a smartphone might present your child. Self-harming videos run rampant on TikTok, from the "skull-breaker challenge" to the "cha-cha slide challenge" (which involves repeatedly swerving a car across a road in time to music). As I write this, videos tagged #passoutchallenge have more than 1.3 million views on TikTok.

These clips are fed to our children swiftly and repetitively by an algorithm that favors the shock factor. How swiftly? A recent report from the nonprofit Center for Countering Digital Hate found that it can take less than three minutes after signing up for a TikTok account before a user is shown content related to suicide and about five more minutes before a user is introduced to a community promoting eating disorder content (Murphy 2022). The research is clear: our children's bedrooms have become crisis centers.

Even if your child doesn't have access to the TikTok app, violent and harmful videos are often shared on YouTube for younger children to find with ease. These videos appear to be innocent, sweet compilations of random acts of kindness or of strangers helping strangers. I found one moments ago after searching for "TikTok beautiful moments" titled "Happiness Is Helping Good Kids TikTok Videos 2022 | A Beautiful Moment in Life #9" (TikTok Fun 2022).

I clicked.

The disturbing video, set to music, depicts a father reading a book. His young daughter—six, maybe—asks for money, and when he refuses, she steals money from his wallet and leaves the house to go on a shopping spree. He finds her in the park with her spoils, is relieved to have found her, then gives her the money she asked for. The clip ends in a hug as the girl winks knowingly at the camera.

On and on, the skits, one after another, escalate in violence and harm. Kidnappings, child abandonment, abusive husbands, cheating spouses, child neglect, rage. In one harrowing scene, a young woman thinks she's witnessing an old woman getting mugged. As the young woman tries to call for help, the old woman puts a chloroform-soaked hankie over the young woman's mouth and she falls to the ground. Dozens more videos just like this one feature this same format—luring music, no voices, real children, all depicting violent imagery: toddlers falling out of windows, kids stuffed in trucks, babies stolen from homeless women.

The account currently has 661,000 subscribers.

Tell me, can we truly rationalize that a smartphone is keeping our children safe? Or is it the source of harm?

Smartphones and Responsibility and Independence

▶ **Convention:** Smartphones lead to responsibility and greater independence.

▶ **Disruption:** What if smartphones don't lead to greater responsibility and independence?

Responsibility

Picture this: Your nine-year-old has been begging for a family dog for months. You and your spouse cave in, and Henry the Golden Retriever arrives. Everyone's smitten, and in the beginning, your daughter takes full ownership: daily walks, cuddles, crate training, the works. And then, things grow a little complicated. Henry's whining keeps your daughter up all night. She's exhausted. Her best friend can't come over anymore because she's allergic to dogs. She's lonely. On neighborhood walks, Henry refuses to stay on the path and keeps trying to lead her to unfamiliar places. She's anxious.

This isn't at all what your nine-year-old pictured. Time after time, unpredictable situations arise, and in taking ownership of a new pet, she realizes the pet owns her. Tell me: would you say your daughter is irresponsible? Or is it simply that she's in a situation in which the costs are greater than the reward, greater than she as a child could possibly have predicted?

Handing our children a smartphone baked with uncontrollable algorithms is like offering them a new dog to care for 24-7. It is not a question of responsibility. It is a question of readiness. More likely than not, you'd rescue your child from a responsibility she can't manage, right? You'd delay the responsibility until she was ready, and in the meantime, you'd carry the burden of care.

Here's the litmus test: If you'd rehome the dog, rehome the smartphone. If you'd personally take over full-time care of the dog, then personally take over full-time care of the smartphone. If you'd stand by and let your daughter struggle her way through—exhausted, lonely, and anxious—all in the name of responsibility? Then, sure, do so with a smartphone. (But proceed with extreme caution.)

Independence

I have the luxury of being smack dab in the middle of the generation that experienced life both before and after smartphones. As a college student, I kept a road atlas in my car. I balanced my checkbook. I used the phone book to schedule a hair appointment, to cancel a meeting, to call the mechanic.

Now I carry a phone that can do all of that for me—and more. Haven't I, then, simply transferred dependence? I no longer need a phone book, a jack, and a phone line to make a dentist appointment. But I do need an internet-service provider, a modem, a router, a power cable, an ethernet cable, a wireless network, network security, a charger, and an outlet. And yes, still, a phone.

Am I truly more independent? Or am I just less dependent on my phone book and more dependent on my smartphone?

A common disruption practice is to brainstorm the many ways progress has, potentially, offered false promises. This logical fallacy is called "appeal to novelty." Are smartphones superior because they're new and modern? And if so, does this automatically count as progress? Once we dissect the false premise that newer must be better, we recognize key oversights in our beliefs about technology's benefits.

For every benefit a smartphone offers, there is, of immeasurable value, a cost. Smartphones make it easier to deposit a check. And also, easier to spend it. Easier to save time; easier to waste it. Easier to send an encouraging text; easier to receive a discouraging one.

And in the case of social media: easier for our kids to find what they're searching for; easier for them to lose themselves in the hunt.

Smartphones and Tech Boundaries

▶ **Convention:** Smartphones teach healthy tech boundaries.

▶ **Disruption:** What if smartphones don't teach healthy tech boundaries?

> *My phone brings me mixed emotions, and I kind of rely on my phone for almost everything. Before I had a phone, my life was so easy and I was a social butterfly, now I'm a seventeen-year-old girl who comes home from work and watches a movie on my phone until my eyes physically will not open.*
>
> —Brooklyn (*Learning Network* 2020)

Boundaries are often best learned through experience. Conventional thinking might lead us to wonder how our children will learn to set their own tech boundaries as adults if they don't have experience to draw from as children, right? To disrupt that idea, let's run a validity check with other scenarios in which legislative boundaries are in place. If we, as a nation, wanted to teach our children better gun control, would we allow them access to their own guns? How about alcohol? Cars? Sex? Pills?

The obvious solution is that if the consequences of engaging in an activity are undesirable or detrimental to the child—or anyone else—we safeguard against it. We protect our child's growing brain and expanding knowledge of the world and we delay permission until a better capacity for wider understanding is possible. The question isn't whether we should teach our children the delicate art of tech boundaries. The question is: Do we do so on a balance beam or on a skyscraper ledge?

There is a difference between taking your son to a shooting range and letting him keep his own semiautomatic on the kitchen counter. There is a difference between giving your daughter a sip of wine at Passover and giving her a bottle to keep on her night-stand. There is a difference between teaching your child to drive at sixteen and giving him a Toyota for Christmas at twelve. There is a difference between handing your child a dose of Tylenol and granting her access to her own medicine cabinet. And there is a difference between educating your child about sex and making him his own Pornhub account. (If this sounds extreme, it's not: 82 percent of sex crimes start on social media, and the average age of first exposure to pornography is now only eleven years old; 93.2 percent of boys and 62.1 percent of girls first see porn before turning eighteen [Gabb 2022].)

⏻ **LOG OFF**

If you have carefully weighed all of the pros and cons and you decide to purchase a device for a child, resist the temptation to give it as a birthday or Christmas gift. Gifts don't come with strings attached, but a personal device does. Instead, offer a device like you would a calculator from a school supply list. "Here's something you might need. Let me show you how it works."

We don't do this with anything else. When we want to teach our kids budgeting, we don't hand them their own credit card and tell them not to spend it after 9:00 p.m. Why not? Because we know the ramifications of financial debt, and we know the long-term consequences of having to crawl your way out of it. So we take a more cautious approach. We start small, in increments.

Slowly, in age-appropriate steps. Chores, cash, potentially a savings account at the bank. We let them do odd jobs. We give them a piggy bank. We take them to the store to make a purchase. And eventually, perhaps when they have their own jobs and incomes, we guide them toward a debit card to use until they're old enough, legally at twenty-one, to decide they'd like a credit card—or not.

The fallacy here is this: we are failing to imagine a future in which our children might *not* want social media. Knowing what we know, why are we helping them navigate something they'd never choose once they're old enough to choose well?

Why are we giving them a dog when we know, without a doubt, that our dogs are driving us all to the brink?

Vision

The central question at this point in the disruption process is simple: What world do we want to be in?

Do we want to live in separate corners of our homes, each lulled into our own worlds by the glow of individual screens? Do we want to spend the short time we have with our kids managing technology limits, monitoring screen time, battling over parental controls? Do we want to run the likely risk that our child will be exposed to something they can't—and shouldn't—yet handle? Do we want to place a wedge between our kids and ourselves, positioning ourselves as tech overlords who grant access to their phones like some sort of behavior-driven candy dispenser?

I don't think we do.

The vision is this: What if a solution exists to disrupt our conventional ways but not dismiss them? What if we could take all of our rationalizations for giving our kids smartphones—to teach healthy tech boundaries, to offer independence and responsibility, to foster social connections, to stay safe in emergency situations—and let them be the guiding force toward a newer, better, truer plan? What if we gave the next generation what we never offered ourselves: the freedom to unapologetically opt out?

We can, and we must.

We can teach tech boundaries through a variety of methods without social media and individual smartphones. We can teach our kids to type. We can teach them to research responsibly, to check facts, to find a recipe. We can show them how to pay a bill. We can let them enjoy a virtual tour of the Louvre, show them how to submit an entry into the local poetry contest, watch a virtual simulation of life on Mars.

We can do all of this—and much, much more—using a desktop computer in the kitchen.

We can teach independence and responsibility through a variety of methods without social media and individual smartphones. We can teach our kids to garden. We can give them dish duty, put them in the laundry rotation, assign them a trash night. We can let them rack up library fines and spend the summer working them off. We can let them call their senators. We can let them write their congressmen. We can let them bike to the grocery store. We can let them paint their bedrooms, design their own skate parks, invent their own desserts.

⊙ DM

"My kid was begging me for a phone, so I told him he could borrow mine for the weekend as long as he used it as a tool, not a toy. He was thrilled! I swiped open the Kroger app and put him to work planning next week's menu. After twenty minutes, he handed it back and asked if he could go ride bikes with his friends."
—Bridget W.

We can do all of this—and much, much more—without a digital footprint.

We can foster social connections through meetups at the park and kickball games and weekly trivia night at a pizza joint. We can show our teens how to negotiate a better car price, how to challenge a point of view, how to apologize. We can teach our kids how to ask for help. We can help them organize a petition. We can let their friends pile into our living rooms, keep our pantries stocked for book clubs and soccer teams and theater troupes.

We can do all of this—and much, much more—without a social-media account.

We can promote safety by teaching our kids how to change a flat, when to speak up, what to do when the power's out. We can ask them to text us when they'll be late. We can show them how to change a lightbulb, change their oil, diffuse a fight. We can teach them to listen to their gut. We can teach them basic first aid. We can show them how to build a fire, how to help a troubled friend, when to give CPR.

We can do all of this—and much, much more—with time, intention, and a flip phone.

A Technicolor Wasteland

Conventional thinking taught us that technology is the future, that adapting is inevitable, that digital literacy is king. "But the research doesn't bear that out," notes *Glow Kids* author Nicholas Kardaras. "In fact, there is not one credible research study that shows that a child exposed to more technology earlier in life has better outcomes than a tech-free kid" (Kardaras 2017).

Still, we trudged forward, and our kids fell in step. We marched to the beat of progress, lured by screens and promises and dystopian dreams. Entranced, we brought to life Apple's 1984 ad, right here, forty years later, in a technicolor wasteland. It's us. We're the faceless, uniform crowd, shuffling along, seduced by technology's empty words and endless manipulations.

But if we should learn anything from the agency that ushered Apple into the spotlight, it's this: a sledgehammer is always close at hand.

✏️ SELFIE

Take a few moments to reflect on the world you'd love your children to grow up in. No boundaries or limitations: What's your sky-high dream? What do you want most for future generations? Peace? Safety? Community? Love? Brainstorm a list of visions and ideals, both large and small. Then on that list, circle what you have the power to influence today.

Chapter 3

The Strategy

Developing a Foolproof Plan

For which of you, desiring to build a tower,
does not first sit down and count the cost,
whether he has enough to complete it?
—Luke 14:28

Mend the part of the world
that is within our reach.
—Clarissa Pinkola Estes

❌ TECH'S PLAYBOOK

> Move fast and break things.

✔️ OUR PLAYBOOK

> Move slow and mend things.

One summer night, the crickets performing a Finzi just outside the open window, my husband, Ken, bounds into the kitchen with purpose. "Let's just bring the whole smartphone thing up before they even want one," he says. "We'll fill in the kids now, have a whole conversation about it. Earlier, the better, right?"

I smile. This is my husband's strategy for everything. *Address it! Head on! Be proactive!* But he's right. We've always made it a point to chat through expectations, milestones, future plans. Why should technology be any different?

So we do. Our daughter is still very young, maybe five. Ken takes her downtown to ride scooters and eat ice cream and feed river ducks. As the evening goes on, he challenges her to look around at people on their phones, to observe what happens next. The next morning, I hear the full recap over eggs.

"Mom, it was *crazy*," she says. "There was this one table where the dad and his kids were all on phones, and the mom was just kind of there. She looked bored, and kind of sad. And then there were, like, little kids, just kind of kicking their feet and yelling at each other while their parents were on their phones. I saw a boy and a girl and I think they were maybe on a date, but they didn't talk to each other? And nobody fed the ducks! *Nobody!*"

Later, I have a conversation with her brother, still a toddler. As I talk animatedly about how phones can get in the way of experiences, I can't help but think that every child psychologist would likely frown on this tactic. *Too much, too soon! So many words! Say less, say less, say less!*

When I finish, his root-beer eyes are fixed on mine. "Mom?" he asks. "Do we have any hotdogs?"

This is how we have grown to become an opt-out family, with the knowledge and awareness that the gift of a smartphone can distract us from the gift of a life. We've strategized a plan from there: our kids can purchase their own phones—smartphone, flip phone, whatever—when they're eighteen, if they want to. It won't be a gift, nor a given. When they drive, a TracFone, a twenty-dollar bill, and an atlas will be kept in the glove compartment for emergencies. Until then, we're willing to install a home phone for their friends to call anytime.

It's a loose strategy, I think. *But at least it's ours.*

Seven years later, it's still in place.

Mr. Zuckerberg's Whims

If you spend enough time in the tech towers of Silicon Valley, you develop a shorthand for startup language. There's a laissez-faire chatter around lobbies, a bit of recklessness on the tip of the tongue. In boardrooms, a collective shrugging of the shoulders. Companywide mottos, like Google's infamous "Don't be evil" (Ghaffary and Kantrowitz 2021), are scribbled on whiteboards to serve as an all-purpose, feel-good salve when things get hairy. Even the taglines developed for tech's newest products are laced with nonchalance (Weiss 2021):

> "When it's this easy, why not?" (HP)
> "Leap ahead." (Intel)
> "Do you dare?" (Microsoft)

Meta founder Mark Zuckerberg himself still adheres to the slapdash strategy he employed when building Facebook from his college dorm room. Explaining one of his core values—which later became Meta's central motto of "Move fast and break things"—a younger Zuckerburg speaks in a college lecture hall, saying, "And a lot of times people are just, like, too careful. I think it's more useful to, like, make things happen, and then, like, apologize later, than it is to make sure you dot all your i's now and then, like, just not get stuff done" (PBS 2018).

And while this haphazard mindset can be helpful in removing the fear of failure that often prohibits innovation, the idea simply doesn't translate when sustaining the result of said innovation. As Meta grew into the global technology conglomerate we now know it to be, employees often complained about "frequent strategy shifts that seem tied to Mr. Zuckerberg's whims rather than a cohesive plan" (Mac, Frenkel, and Roose 2022). When Meta moves fast, we— and everyone else on the rollercoaster—follow. And when they break things? We all fall.

Meta's long history of breaking things—both purposefully or by sheer oversight—is exhaustive, including Cambridge Analytica's mass collection of data and invasion of privacy, possible Russian interference in the 2016 presidential election, unrestrained hate speech, incitement of the Myanmar genocide, the viral spread of disinformation, and more (Sadowski 2021). "Despite the executive team's awareness of these serious problems," writes technology researcher Jathan Sadowski, "despite congressional hearings and scripted pledges to do better, despite Zuckerberg's grandiose mission statements that change with the tides of public pressure, [Meta] continues to shrug off the great responsibility that comes with the great power and wealth it has accumulated."

But we know better. The Move Fast and Break Things era is over, at least for our families. We, as parents, have the power—and the responsibility—to build the safe, sustainable culture that Silicon Valley has continually failed to create—for our families, for our communities, and for our world.

How? For starters, we take tech's formula and do the opposite.

Move Slow and Mend Things

Moving slowly and mending things, as a philosophy, is symbiotic in a sense. To truly mend things, you must move slowly. And if you're moving slowly—thoughtfully, with care—chances are you'll have fewer things to mend.

Author and slow-living advocate Carl Honoré describes this

delicate balance: "Children need to strive and struggle and stretch themselves, but that does not mean childhood should be a race," he writes (Belkin 2009). "Slow parents understand that childrearing should not be a cross between a competitive sport and product-development. It is not a project; it's a journey."

His advice requires a healthy mix of freedom and intention, giving children plenty of time and space to explore the world on their own terms, keeping family schedules as open as possible, limiting options, rediscovering simple pleasures, allowing our children to work out who they are, rather than what we want them to be.

His wisdom is sound. And in 2020, many of us found out why.

LOG OFF

Visit an antique store as a family and point out an item from a previous era, such as a washboard, manual pencil sharpener, or typewriter. Let your child guess what these objects might have been used for in "the good old days" and how they shaped the lives of your ancestors. Then ask a beloved elder in your life if they have memories of using that type of object. It's never too early to teach that our environment has been directly influenced by the ideals and innovations of the past—and why innovations require both responsibility and care.

Whatever's Fastest

The COVID pandemic brought much turmoil—closed businesses, uncertain futures, distanced relationships—but it also brought a welcome surprise: clarity. With a jolt, we were forced to examine who we were without our schedules, our habits, our lives beyond our four walls. Families collectively slammed the brakes on soccer and ballet, on Lego club and library runs, on HIIT at the Y and weekend concerts downtown. Parents recalibrated schedules, shifted childcare, figured out how to get toilet paper or do without.

And, lacking a better plan, schools funneled kids into a crash-course remote-learning system that, later, NPR reported brought "unprecedented, historic learning loss" (Turner 2022), which *The Atlantic* admits "is far greater than most educators and parents seem to realize" (Kane 2022). Shutting down our schools, says *The Economist,* "was worse than almost anyone expected" (Economist 2022).

We moved fast, and we broke things.

I witnessed the shattering firsthand. Shortly after US schools shut down, I began guiding quarantined families through an alternative to remote learning: a screen-free early childhood program, Other Goose. Daily, I communicated with caregivers seeking reassurance that their child's education would stay on track. But when pressed for which "track" they were referring to, for what standards they'd like to see their children stand for, the majority of parents shrugged their shoulders. One response? "Whatever's fastest."

I watched parents bend over backward to juggle their daily conference calls and their child's Zoom classroom schedule, and I witnessed burnout across the board. One parent asked if I could recommend a Skype math tutor for her two-year-old. Another asked what sort of STEM programs were available for babies. Time and time again, the questions were fired off, laced with the deepest fear a parent knows: *Help! My kid's falling behind!*

It became clear to me that in the absence of a sound strategy and the confidence required to implement one, we were all susceptible to Meta's harmful "Move fast and break things" mentality. We need thought, and we need the time and

⏻ LOG OFF

Offer your child a forward-thinking mindset with this single sentence: What's another way? When your toddler's block tower falls, ask, What's another way your plan might work? When you're microwaving chicken nuggets, ask your kids, What's another way I could heat these? When your teen can't borrow the car, ask, What's another way you could get there? When you're texting a friend, ask, What's another way I could keep in touch with this person? The practice opens our minds to alternative solutions and forces us to rethink our long-held habits, patterns, and beliefs. This week, tape "What's another way?" to your fridge as a reminder to try this phrase in your daily rhythm.

agency required to think it. So in my home I began to refer to our quarantine as a gap year, and I encouraged the families inside Other Goose's program to take a wild, audacious pause. To chuck expectations and standards out the window and to let the year unfold as an education in something greater than times tables. Our focus would not be on quick academic wins, hours logged in on Zoom, and busy work in the form of ABC coloring sheets. Instead, we would take walks. We would count raisins. We would build forts. We would borrow a new board game, sing an old song, read Dickens.

We weren't going for fast. We were going for *far*.

A Gap in the War

The practice of a gap year originated in the 1960s with the intention of delaying academics to teach young people self-confidence, practical awareness, and community care. In other words, to move slow and mend things. It was one strategy of many, a way of giving an entire generation a longer runway to seek respite from the severity of their parents' wars. And six decades later, our young kids were seeking the same: respite from the severity of our many, many battles. Why not grant them a gap in the war?

In the United States, this sort of pause might easily be labeled as frivolous. Anything other than full-speed ahead is a life of leisure, yes? For many, opting out of remote learning might appear to be a privilege. But opting in? Ushering in a veiled education where nearly 90 percent of remote-learning tools are designed to send a child's data to third-party companies (Harwell 2022)? Adapting to a digital culture where children are exposed to Zoomwide cyberbullying and hate speech (Cherry 2020)? Where kids are required to increase their screen time—up to six hours per day—in the name of better test scores (Prothero 2023)? Where Google secretly and unlawfully monitors and profiles children through their assigned classroom tools, forcing children to surrender their biometric data to third parties with no assurance of its security (Mühlberg 2020)?

For some, opting out will be the only way to survive.

Tell me: does that make it a privilege, or a right?

George Washington for Ona Judge

The idea took root, and I penned an op-ed for the *New York Times* encouraging more parents to reclaim their agency. Our nation's young children don't need more remote learning online, I argued. They need more runway to learn—offline. Week after week, I heard from parents who grew in confidence and began to build a tech-free family culture of their own. Clear values emerged: Kids learned to display character over compliance. Parents were granted the freedom to swap out Huck Finn for the Downstairs Girl or George Washington for Ona Judge. Uniforms were shed, and so was uniformity. Families abandoned Zoom science class and spent more time watching clouds. Teens became part of the family life and home structure, whether through accounting spreadsheets or blueprint renderings or sauteed garlic. Preschoolers gained independence. Kindergartners made their own lunches, washed their own socks.

We maneuvered IKEA desks into bedroom corners because the light was better, they're farther away from the drumset, and there's a bookcase nearby. We tackled the wallpapering project. We fixed the deck. We finally splurged on that backyard trampoline. (And thank goodness for *that!*) We started practicing a weekly Sabbath dinner, beeswax candles and all. We played board games over lunch. We taught ourselves to make sourdough. We invented new traditions, like an Easter egg hunt on the first Tuesday in July, because we were hot and we were bored and we could.

We fought. We cried. We moved slowly.

We mended things.

A Sustainable, Renewable Strategy

Now, years later, many of us speak of the hardship of that pandemic in ways we're only beginning to process. And yet, when pressed, we talk of having more difficulty in the months that followed the reopening of our country, of how hard it was to reenter a society that had been permanently altered. Everything seemed fast and loud and big. Our social mannerisms and nuances had shifted. Our small talk was rusty. Heck, we'd stopped wearing bras.

We were confronted with a new dilemma. Everything we once knew had been stripped away, for a time. What, then, did we want to add back? And how? Could we ensure that, as Rebecca Solnit so aptly writes, our new lives do not move "faster than the speed of thought, or thoughtfulness" (Solnit 2001)?

Some of us chose to hop right back into our "old selves," happy to be comforted by familiar routines and rhythms. Others radically altered their lives by making cross-country moves or massive relationship shifts, or trying a new educational method for their kids. The rest of us, I imagine, are still figuring it out and course-correcting as we go.

And that's the point. We are not blindly following some predetermined path, an algorithm laid out before us that ends at a destination no one truly understands. We are selecting our paths. We are choosing our plans. We are opting in on the things that matter to us, and opting out of the things that don't. And wherever we land, we have the luxury of employing a sustainable, renewable strategy that Silicon Valley never could (or would).

We must only ask ourselves these questions: What feels too fast? What feels broken? How can we slowly mend?

Birds, Not Beeps

The answers to these questions might involve massive, mountain-moving shifts. A family we know felt as if they were drowning in their fast-paced, plugged-in LA lives, so they put their house up for sale, pulled the kids out of school, and traveled the country in a van to "town shop" for a new place to live. They stopped to visit friends, to try on different lifestyles, to see what fit best. Along the way, every adventure brought them closer to each other, and closer to a new family rhythm. They lit fireworks, danced in a downpour, tried deep-fried Oreos at a small midwestern fair. After a long season of searching, they chose to make their home in a quaint fixer-upper in Bend, Oregon, joyfully settling into a life that was the opposite of all they'd once known.

Most families will find that a "Move slow and mend things" strategy won't involve sweeping changes and that a single ripple of change can refresh an entire day. Earlier this year, my friend's daughter was beginning kindergarten, and the transition was not going smoothly. "All of a sudden, it just felt like life was happening really fast. The mornings were rushed to get dressed and sign the papers and make the eggs, and then it's like, 'Hurry up and get in the car! We're going to be late!'" My friend paused, talked herself through which parts of the day she could perhaps slow down a bit. "And it hit me! I felt rushed because I *was* rushed! There's this sign in the carpool lane that says, 'Drop, don't stop,' you know, to keep the line moving. But it flashes this red light, and if you don't go through fast enough, it starts beeping at you, and there's, like, no time for goodbyes or hugs or any of it."

So she started parking a few blocks away at a coffee shop, then walking her daughter to school— slowly. They jump over sidewalk cracks, look for caterpillars, hum

🔄 **OPT OUT**

A simple way to serve the "Move slow and mend things" strategy is to ask, What did people do before [blank]? Before Facebook Marketplace, there was Craigslist. Before Twitter, there were newspapers and bulletin boards. Before Apple Watches, there were pedometers. Peer back five, ten, fifteen, or even twenty-five years ago to reintroduce yourself to ways that might better your life without the harms of an addictive algorithm or persuasive technology.

little tunes to each other. "We hear birds, not beeps. It's just the most delightful part of my day. And it's a small thing, sure, but slowing down that one transition made all the difference in how we thought about the school day."

Everyone Had Them

The beauty, of course, is that we're all in it together. All across the world, communities and movements and groups are cropping up as a commitment not only to move slowly but to move slowly *together*.

Brooke Shannon, leader of the nationwide smartphone-free movement Wait Until 8th, believes in the power of mending collectively. Brooke tells me the idea for the movement was sparked after a group of parents from her daughter's elementary school in Austin were discussing mounting pressure to give children their own smartphones at an early age. "We started seeing children as young as first and second grade coming to school, play dates, and birthday parties with the latest iPhone," she says. "As we started to ask around, many parents said they eventually caved on the smartphone because 'everyone had them' and they did not want their child to feel left out."

So Brooke and her fellow parents rallied together. "We designed the pledge to flip the script on peer pressure and use group momentum for good," she tells me. "We made it this way so you don't have to fear 'What if I am the only parent in my kid's grade that signs this?' This takes the pressure off; the pledge kicks in only when at least ten families from your grade sign it."

Now more than fifty thousand families from all fifty states have signed the pledge to delay giving their kids a smartphone until at least eighth grade, and the number keeps growing. These are kids who will climb trees and wade in streams and fail far outside of the public eye, kids who will be protected from the weighty responsibility of owning a device that owns them, kids who will be well on their way to building a vibrant life full of experiences, one that offers what the world's burgeoning platforms have sold but can never deliver.

Whether we are delaying smartphone use until eighth grade or far beyond, the beauty of the "Move slow and mend things" strategy is that, at its core, it is us. Ourselves. Our own families. Our quirks and tremors and loves. This is not an expert strategy devised from the depths of a college dorm room or the heights of a Cupertino tower. It is not fed to us by an algorithm. It is our own, derived from our actual lives, using what we have, not what we *want* to have, seeing where we are now, not where we think we should be by now. This approach asks, What have we been given? What is here, right in front of us, that we can make just a tiny bit better? And where might that lead next?

Moving slowly and mending things invites connection, elicits cooperation, and ripples far beyond our homes and families. Over time, we'll find ourselves looking a crossing guard in the eye, thanking him for keeping our kids safe. We'll find ourselves making the time to help the grandfather in aisle 9 find green olives. We'll find ourselves pausing to feed the ducks.

"We must make the conscious decision to just . . . pause," agrees Kim John Payne, author of *Simplicity Parenting*. "I don't think there's been a point in any time in history that I can think of— and I challenged myself to think of one—that is as crucial as a parent's decision around screens. We get to make this decision that so affects our children's future, so affects their brain development, so affects their social and emotional abilities. War happens and terrible things happen, and we don't get to decide that. But this? This undeclared war on childhood? We get to decide what to do, if we want to."

And we must.

> ✏ **SELFIE**
>
> Take a few moments to reflect on the world you'd love your children to grow up in. No boundaries or limitations: What's your sky-high dream? What do you want most for future generations? Peace? Safety? Community? Love? Brainstorm a list of visions and ideals, both large and small. Then on that list, circle what you have the power to influence today.

Part 2

Chapter 4

N of 1

Discovering What Delights Our Kids Most

God only knows what it's doing to our children's brains.
—Former Facebook president Sean Parker

Why, even the hairs of your head are all numbered.
—Luke 12:7

❌ TECH'S PLAYBOOK

> ⬜ Find out what people like so we can fill our pockets.

✅ OUR PLAYBOOK

> ⬛ Find out what people love so you can fill their hearts.

We are driving to a friend's pool party when I ask Ken to stop the car. I have seen something on the side of the road, left abandoned next to a mounting trash heap. It is dirty. It is broken. It is the perfect birthday gift for our two-year-old.

"Can you load this into the trunk?" I ask him.

"An old kitchen sink?" he says. "We don't have space for it."

But he does it anyway, and that is how our daughter spends a summer splashing around in a custom (free) water table just her size, with a soap dispenser for bubbles, no less.

I am thinking of this story when I'm on the phone with Kim John Payne, author of *Simplicity Parenting*. He is calling from his farm in New England, and as he speaks, I catch his gentle humility betraying the internationally acclaimed and highly sought-after Waldorf educator that he is.

I have called him to ask about engaging our kids, about providing them with a low-tech experience and ensuring it doesn't scar them, or at the very least cause them to lock us away in a nursing home far earlier than we'd like. We speak of his grown daughters, how they are opt-out kids, too, how they are thriving in this world despite—or because of—a childhood spent without screens. He walks me through the intentionally addictive nature of the video game Minecraft, which is often dubbed "digital heroin" after recent brain-imaging research shows how the game affects a brain's frontal cortex—which controls executive functioning, including impulse control—in exactly the same way that cocaine does (Kardaras 2016). "If there is such a thing," Kim jokes, "those [game] designers are going to have a very hard time at the pearly gates!"

He notes with compassion how difficult it is to parent today without the use of screens, but recognizes that it is far more difficult to parent *with* them. "A child in front of screens is receiving continuous dopamine, this pleasure and reward feedback cycle, and it's constant, and [the child is] being trained to seek it out," he says. "And if you tell that child, *Okay, it's time to clean up the table now because supper is being served,* well, that's not pleasurable and it's not rewarding."

Kim tells me that parents all around the world are discovering that when they lower interactions with screens, their children are much more flexible. "It doesn't mean kids won't still not want to do stuff, you know? But you can work through it. You don't meet this snarling beast when you ask a child to do the simplest things."

And then he tells me what he has recently learned by eavesdropping on a children's marketing conference—one that packages delight and sells it in the form of cereal, cars, toys. He says that his interest was piqued by visiting one particular session, attended by several hundred people, titled "The Removal of Purchasing Friction."

Kim says that this small segment of children's marketing and advertising accounts for roughly a $16 billion industry. I am nodding along, thinking of the many ways technology has funneled money into the removal of any friction from our lives at all—one-click purchasing, infinite scroll, autoplay, personalized recommendations—that I almost miss what he says next.

"We know what this industry means by *removal of,*" he tells me. "But what do you think they mean by *purchasing friction?*"

"Choice," is my guess.

"Parents," is the answer.

In panels and forums and handouts, attendees were taught dozens of tricks for "removing purchasing friction, aka parents." Over and over, the two terms were used interchangeably. It seems, to the world of tech-based advertising experts, we are not our child's guide. We are not our child's protector. We are not their confidante or companion or mentor.

We are, quite simply, in the way.

And the message, backed by $16 billion dollars, is this: Your parents can never delight you. But we can.

It's Like Magic

Ask any TikTok user what they love about the app and they're bound to use the word *new*. "I can't even tell you how many new things I learn about from here," said one teen I talked to. "The people I follow on TikTok just get me. They're sharing, like, all this stuff I never knew I needed to know, all these things I'm obsessed with now."

I asked her to show me her feed. Upon opening the app, I was welcomed by a conglomeration of her entire personality: plant care, makeup tutorials, piercings, '80s music covers. "I'm recently getting really into Bowie," she tells me, shortly before I spot the iconic *Labyrinth* movie poster on her For You page.

If what our kids want is people who get them, TikTok is poised for gold.

TikTok's feed is similar to other social-media apps, but the emphasis is on finding new ideas, products, and trends. It's a discovery platform. As soon as you open the app, you're not met with people you're following, you're met with people you're not following *yet*.

TikTok features prominently its For You page, a curated blend of popular user-uploaded videos—looped videos, all less than a minute long. Total strangers dancing, goofing off, trying pranks, challenges, recipes. What you like, you get more of. And TikTok knows exactly what you like. "It's crazy," says a TikTok user. "I'll get a craving for, like, cereal, and the very next video is this cool chick eating cereal. It's like magic!"

It's not. In addition to traditional metrics (people you follow, where you live, what your hobbies are) TikTok relies on biometrics to inform its algorithm (Wouters and Paterson 2021). While you're watching TikTok, it's watching you. Through the camera on your phone, TikTok can detect changes in your face as you react to each and every video you watch. What makes you smile? What makes you laugh? What raises your eyebrows? And for how long? TikTok studies that information to determine what you'll like next—and better.

Then they use that information to display ads from its brand partners, who—primed with your most intimate data imaginable—can guide you down whatever path will guarantee them the most profit (Chiara 2023). Imagine: you're a teenage girl watching a typical TikTok dance challenge. But through biometrics, the app registers that your eyes are distracted by the flat belly of one of the dancers, her toned thighs. In the next video, you're zeroing in on someone's triceps, and your heart beats faster. The next video? An influencer in an undisclosed ad for a three-day detox cleanse and workout ebook to guarantee "you'll look amazing for your next TikTok." Just $199.

Click. Add to cart.

⏻ **LOG OFF**

When's the last time you gave your child your full attention and focus? Eye contact builds warmth, trust, and connection. Take a cue from TikTok and host a staring contest with your child—not for data but for delight.

You needn't imagine. The technology is already here.

Aside from amassing biometrics to boost their bottom line, what else can TikTok—and their third-party trackers—do with that information? That's often the primary concern cited by many lawmakers and political advocacy groups. Because TikTok, which boasts more than one billion monthly active users, is owned by Chinese company ByteDance, the FBI alleges the app poses national security risks (Treisman 2022). Under TikTok's privacy policy, which enables the app to collect both "faceprints and voiceprints" from any user in the US (Huddleston Jr. 2022), the possibilities—and harms—are endless.

But for parents (ahem, purchasing friction), the concern hits closer to home. TikTok's biometrics aren't used just to sell your kids products.

In some cases, they'll be used to sell *your kids*.

She Was Mortified

Ever seen a deepfake? A deepfake is a video that employs AI, or deep learning, to copy someone's face, mimic their hand gestures, learn their movements, and borrow their voice to impersonate them in videos. If you've ever asked yourself why you just saw Biden sing "Baby Shark" or Tom Cruise selling toothpaste, you've seen a deepfake. For most users, deepfakes are harmless ways to channel their creativity and stretch the limits of what we can do with technology. But with the skyrocketing capabilities and widespread utilization of AI, they're increasingly being deployed in cases of identity fraud, extortion, and disinformation campaigns.

And worse. According to research from the AI firm Sensity, a shocking 96 percent of deepfakes online are pornographic and used to target girls (Dickson 2020). One mother of a teen found that one of her teen's TikTok videos had been deepfaked and shared on Pornhub. "She was mortified. She didn't want to go back to school," her mother said. "It's not a lesson you should have to learn at seventeen."

While several apps and software are readily available to make deepfakes easy for anyone to create, AI impersonators insist that the art hasn't yet been perfected (Fisher 2022). "The most difficult thing is making it look alive," says VFX specialist Chris Ume. "You can see it in the eyes when it's not right" (Vincent 2021).

But don't worry, TikTok has a fix for that. There's the "pupil challenge," which encourages users to upload zoomed-in videos of their eyes while they think of their crush (Pilkington 2022). "If your pupil dilates," says one user, "it's true love!" (Musiart 2021). And the "side eye challenge" (Smith 2021), where TikTok users watch in awe as their eyes change color. "To take part in the trend," reports PopBuzz, "all you need to do is hold your camera as close to one of your eyes as possible and then look to the side on the beat of the music" (Smith 2021). And then, of course, 2020's "eye color challenge," where users "recorded their eyes while keeping the camera flash on and then used TikTok's S5 filter to find out their 'real' eye color."

If better data is what they want, better data is what they'll get.

In fact, TikTok recently debuted a new feature called TikTok Now, introduced as a way to offer "deeper connection and entertainment in a fun format." How? By inviting each user to "capture what you're doing in the moment using your device's front and back camera" (TikTok 2022).

Better data is *exactly* what TikTok wants.

His Favorite Memory

So how well do you know your kids? If a $50 billion algorithm can combine all of your child's passions and interests—both conscious and subconscious—and spit out something engaging in just a few seconds, are we a match for that sort of machine?

Emphatically, resoundingly yes.

Successful algorithms rely on a strategy called predictive personalization. The program takes just a few pieces of data—location, age, interests—and uses that information to predict someone's behavior, needs, or wants to "precisely tailor offers, products, and messages to each recipient across channels and touchpoints" (Listrak, n.d.). This is why your Instagram ads are continually showing you products you didn't even know existed but are exactly what you've been looking for (ahem, weighted blankets). This is why a teen who's just getting into Bowie's music will binge-watch *Labyrinth* tonight while her best friend pours herself another bowl of cereal.

If technology specializes in predictive personalization, let's be parents who specialize in predictive *humanization*. Let's protect our children from anyone—or any place—that seeks to manipulate their interests for profit or gain or worth. Let's, instead, guide them to places that offer delight while asking nothing from them in return: a bed of pine needles, a crackling hearth, an old kitchen sink turned water table. Laps for reading, skies for gazing, hammocks for swaying and dreaming. Delight.

No matter what your family dynamic is right now, you have this very tool at your disposal. You, as a parent, have the potential to employ this strategy better than any AI machine-learning algorithm ever could. Why? Because TikTok can't tell your daughter how she liked her strawberries cut when she was three. TikTok doesn't have data for your son's first words or his favorite memory or the name of the stuffed teddy that he slept with for eight years straight.

But you do, and it's all you need to start. You have the data. It's time to put it to work.

⏻ LOG OFF

Screens may offer children a lot of sensory experiences, but the one thing tech can't (yet) replicate is this: touch. A shoulder squeeze, hair-braiding session, thumb war, simple hug, or fist bump can go a long way in delighting your child. Start small, but work your way up to a baseline that feels good to everyone in your family. As author and family therapist Virginia Satir once said, "We need four hugs a day for survival. We need eight hugs a day for maintenance. We need twelve hugs a day for growth" (Comaford 2020).

Study Your Child

Take some time this week to watch your child. What kinds of topics do they bring up naturally? How do they react to something their sibling said or did? What do you notice that makes them smile? Laugh? Are they particularly animated when they talk about someone they know? Do they stand up straighter when they perform a certain task or when they receive good news? How are they handling conflict? What do they do when they're overwhelmed? Where do they retreat to, and what do they do next?

Pay special attention to micromovements, gut reactions, and "tells." Does their smile reach all the way up to their eyes? Do they giggle when embarrassed? Shut down when frustrated?

The truth is, we often paint a picture of our child that's static, finite. We remember them clearly at a certain age, maybe from a key vacation or a time we felt most connected. And with the distractions of daily life, we forget to recalibrate that picture. We fail to look at our child as they are *today*, and we revert to the image in our minds of who they once were.

🟢 **+FOLLOW**

Consider reading Gary Chapman's bestselling book *The Five Love Languages of Children* for a jump start to learning better what makes your child come alive.

So today, recalibrate. As you study your child, stay open minded and curious. Resist the temptation to step in with a solution or fix anything you're witnessing. Just watch and learn. This isn't about behavior management. It's about data collection. Remember: the algorithm doesn't process only desirable results. It processes *every* result. Think like a scientist and gather your data.

Practice Strewing

Once you've studied your child, themes are bound to arise. Your daughter lights up when she sees the neighborhood dog, and your son can't stop talking about the martial-arts birthday party he went to last week. Your toddler is newly enamored of her older brother's velcro sneakers.

Here is where your work begins. Ever heard of strewing? It's a tried-and-true educational method rooted in the idea that a child cares more about an idea when they discover it independently or stumble upon it accidentally. It's tremendously engaging and surprisingly hands off for the parent. In strewing, you simply lay out an assortment of things for a child to discover independently. Not too many, and not all at once, but enough for your child to notice within the home. Perhaps you borrow a *Benji* Blu-ray from the library and leave it on the coffee table. Maybe you pull out your husband's old tae kwon do nunchucks from the attic and leave them on the kitchen counter. Can you find some velcro in your old sewing kit for your toddler to enjoy?

Just as their eyes would light up on their TikTok discovery page, their natural curiosity will be piqued. From here, offer them the independence to explore, offering age-appropriate guidance along the way.

Invite a Deep Dive

As the newness of this discovery wears off, you'll begin to understand a clearer picture of what your child's interest is rooted in. Take your time to gather more information. Maybe *Benji* was a flop because your daughter didn't actually light up over the neighborhood dog itself but over the fact that the neighborhood dog happened to be wearing a sweater, and how cute was that? And your son liked the ninja party because he wore a mask but couldn't care less about nunchucks. Turns out your toddler ignored the velcro from your sewing kit but loved organizing your many-colored threads. In the process of deep-diving, you'll learn the nuances of his or her interest so you can begin the strewing process again with clearer, more targeted information—just like the algorithm does.

Establish Check-Ins

Once you've successfully landed on your child's interest, help them solidify this interest as part of your family rhythm. You can take this any direction you'd like! It could be as simple as taking daily walks to visit her favorite sweatered dog, or as dedicated as teaching your daughter how to knit a new dog sweater for him for Christmas. Whatever you decide, help build momentum for your children's interests by encouraging them toward a particular goal the whole family can help support.

Of course, interests come and go quickly in a child's life, so feel free to leap around and delve into new ideas. No data will be lost, no experience will be wasted. Remember: think like a feed! Something new and exciting is just waiting to be discovered—for your child, for you, and for the whole family.

Ask Questions

A popular way TikTok creators elicit engagement is by posing a question that sparks instant curiosity. Did you know a banana could do this? How come no one told me I'm doing my mascara the wrong way? Employ the strategy with your child, inviting him or her into exploration with you.

Discovery is a two-way street, and our kids love to see us learning right alongside them. The next time you learn how to soothe a paper cut with lip balm or cut a watermelon with dental floss, call your child over to see. Once they witness you embracing a teachable spirit, they're more likely to approach you with new things they're learning, too. The result? A home that flows freely with ideas, inspiration, and shared experience.

Make a Prediction

By now, you'll have mentally and informally gathered dozens—if not hundreds!—of data points between you and your child. It's time to employ predictive personalization: if your child likes X, will she then like Y? While it sounds complex, I'm willing to bet you employ this strategy nearly every day. If my son likes tomatoes, will he eat this curry? If my teen likes hikes, should we all go backpacking this summer? If my baby isn't sleeping well, should we try a sound machine? Blackout blinds? Earlier bedtime?

Here's where you're guaranteed to beat the algorithm. You're using a foolproof strategy: an N of 1. An N of 1 trial is a case study done on a single patient to find subtle results that get lost when employed on a larger

> ⬇ **DM**
>
> "For Christmas one year, my daughter asked for mugs for her new apartment. She received three of the exact same one, and we quickly found out everyone in the family had discovered it through Instagram ads. It took the fun out of it. Even though the mugs were perfect for her, they felt less special because it didn't seem like a genuine gift that had taken time or thought. Recently, my husband and I dug through my mother's attic and found her an old goofy mug from the '50s. She loves it. She drinks out of it every day."
>
> —Mollie D.

scale. "Many measurements from one person over a period of time rather than fewer measurements from a large pool of people lead to specialized optimization" (Schmoon, n.d.).

And that's where you, the parent, can shine. You can specialize. You can optimize. You can look at every ounce of data and, you, the expert analyst, get to decide what happens next.

Better Results Tomorrow

Your predictions aren't going to be perfect. But the algorithm isn't either. Maybe your teen wrinkles his nose at your attempt to recreate his favorite falafel order. Maybe your daughter hates the book you brought home from the library. Maybe your son is less than impressed with your magic trick. Don't fret. That's the point. You've created another data touchpoint, one of many you can use to build on in the future.

And if we've learned anything from the social-media giants so far, it's this: more data today means better results tomorrow.

⊘ SELFIE

How well do you know your kids? Journal through your experience working through predictive humanization with your child. At the top of a blank sheet of paper, list each child in your family. Underneath each name, make a list of recent moments that brought them joy and laughter. What sparked the delight? How can you recreate a similar moment? What will you try next? Get as specific as possible. Then give your favorite idea a go! Report back to your list often to discover trends or areas worth revisiting. It's all data, and it all matters.

Remember: you don't have to rely solely on the data collected between you and your child. Look around at the experiences of other families you know and respect. Research what's happening in the lives of today's younger generations. What patterns are arising? What are you noticing? Take a few moments to think about what you're seeing and what it might mean for your family.

Chapter 5

iPerson

Supporting a Low-Tech,
High-Impact Educational Experience

Education is the movement from darkness to light.
—**Allan Bloom**

See to it, then, that the light within you is not darkness.
—**Luke 11:35 NIV**

⊗ TECH'S PLAYBOOK

 May a child's mind be guided by machines.

✓ OUR PLAYBOOK

 May a child's heart be guided by discovery.

This afternoon, I took a walk through the woods surrounding our neighborhood school. It is a small school, fewer than five hundred children, and, nearly a decade ago, was often touted for its excellent test scores, warm and approachable teachers, and outstanding parental involvement. On the outside, it looks very much like the public school I remember attending as a child. There's a clanging flagpole, rusted basketball hoops on the playground, a single-file army of yellow school buses awaiting the closing bell.

But as I pass along the sidewalk and glance inside the windows, I see a marked difference from the classrooms we all once knew. There are no clay pots lining the windowsills with eggshell seedlings. There are no finger-paint tulips taped along the door frames. In fact, I pause to register any signs of spring before I'm distracted by an oversized TV screen in the center of the room. In bright, flashing lights, the TV is displaying a countdown: 4:38, 4:37, 4:36 . . . The children's heads are all hunched over their individual screens, oblivious to the marching orders of the clock. I watch as the time flashes onward, looking for clues as to what happens when it hits zero.

Later, a friend tells me this was her child's recess. The day had been "too cold to go outside," so the kids were granted extra iPad time.

"Just educational apps, though," she assured me.

Irreparable Harm

Many parents were first introduced to these educational apps during the COVID-quarantine remote-learning experiment. Learning apps like HOMER, Kahoot! and Prodigy Math were assigned as homework to upward of ninety thousand schools throughout the nation (Fairplay, n.d.). But according to childhood advocacy group Fairplay, irreparable harm is baked into these apps. From promoting excessive screen time to deceiving parents and educators, many of these apps teach our children to be consumers, not learners.

"Prodigy may keep children quiet and happy while teachers or parents are busy, but it doesn't teach them math," says Faith Boninger, a commercialism-in-education researcher at the National Education Policy Center for University of Colorado Boulder. "Research indicates that kids must spend hours in the game to improve their math achievement scores by just one point. That might not be so terrible, perhaps, but during those hours they endure emotionally abusive marketing until they convince their parents to shell out money for a membership. Under a pretense of teaching math, Prodigy is using schools to access and manipulate a lucrative child market" (Fairplay 2021).

Kathryn Starke, a national literacy consultant, agrees that personalized learning is failing our kids—in more ways than one. "I have seen the impact of tech in classrooms, especially in the primary grades," she tells me. "This higher iPad use is negatively affecting student behavior in stamina of reading traditional text and writing. Problem-solving skills, critical-thinking skills, collaboration, creativity, and attention spans of young children have [all] seen a decline since this increase of technology in the K–2 classroom. Unless students share the details of their educational day at home, I'm not sure all parents are aware of the amount of time their children are experiencing screen time in elementary schools on a daily basis."

For many parents, these assigned learning apps were the gateway into a digital future they didn't envision for their kids. "I was all for personalized learning when my kids' teachers started talking about it," writes one parent. "But I didn't realize how much screen time it would mean, and how much homework would be assigned on a device. I can already tell [my fifth grader] is becoming addicted to the rewards of these learning apps. It's really created a wedge between us. I feel like I'm standing between his success at school whenever I place limits on the device."

If parents are having a difficult time placing boundaries on their children's app use, it's no wonder teachers are feeling just as powerless. "The administration really pushes for better test scores, which often means more content input," noted an elementary teacher I spoke with. "But I'm seeing the negative effects on our kids every day; there are growing outbursts, skyrocketing distractions. I want to just pitch the iPads in the garbage and start over, go back to the way learning used to be—the reasons I went into teaching in the first place."

Starke agrees. "I have been in education for almost twenty years. . . . An iPad-centered classroom experience is certainly not what I envisioned when I set out to work in education," she says. "School districts around the country often have a budget that is allocated strictly to EdTech, which means schools have to implement various apps and platforms, whether or not the options on the market are beneficial."

If anyone might see the value in higher budgets for tech, it would certainly be those in the tech industry, right? The founders, the programmers, the futurists? No. Many of tech's most prominent leaders advocate for the use of less technology in the education of their own children (Akhtar and Ward 2020).

Wired founder and father of three Kevin Kelly has a theory why (Haydon 2018). "I'm not sure we want the environment of school to reflect only the current media landscape," he writes. "If you have an education that mirrors society too much, it won't be effective in improving it. There is value in quiet contemplation, going deep as you might do into a book, being involved in nature. . . . As we remodel education, the target is not to produce better workers but wider thinkers."

And yet many classrooms simply aren't equipped to adequately address a wider-thinking learning experience, and our national test scores, coupled with our growing rates of childhood anxiety and depression, reflect that. The barriers are many: administration, access to nature, classroom sizes, budgets, and, of course, technology. Says one seven-year-old I spoke with, "My teacher doesn't like us to play in the mud at recess because it makes the iPads dirty."

Many educators admit that their hands are tied. "We've got to guard those iPads with our lives," one grimaced. "You know, because of the 1:1 program."

I didn't know, not exactly, so I contacted someone who did.

Every Student's Backpack

I arranged an interview with Darrell Lynn, a name you might not be familiar with, but a name Apple's founder, Steve Jobs, certainly was.

In 2002, Lynn, a director at Apple, worked alongside Steve Jobs to establish Apple's Lighthouse Project, which aimed to put iBooks into every student's backpack (Sellers 2002). According to Lynn's PR representative, as a result of his efforts, Apple's share in the education market grew from 11 percent to 40 percent. He later created and led the company's 1:1 program, shifting the deployment of computers in education to a more consumer-facing model, increasing the company's education market share to a dominating 63 percent.

If I was looking for the guy who put an iPad in my child's backpack, I'd found him. I communicated with his PR specialist, who was happy to help round out my research on this chapter. Until I touched on one subject in particular.

My question to Lynn was simple: "Knowing Steve Jobs didn't allow his kids to have iPads, how did [Lynn's] team navigate the ethics of this initiative?"

Months later, I'm still awaiting a response.

An iPerson

More cords, more plugs, more algorithms, more webs, more input. Silicon Valley calls this "personalized learning." In the world of K–12 educational technology, personalized learning is often defined as the utilization of digital devices "to tailor instruction to each student's strengths and weaknesses, interests and preferences, and optimal pace of learning" (Herold 2016).

Lovely in theory, right? But it doesn't add up. If we're tailoring our educational standards to reflect each child's pace, why are we still testing for nationwide averages? If we're accounting for every child's interests and preferences, what's to be said of those who prefer a low-tech educational experience? Where's the app for that?

Personalized learning is little more than a veiled strategy to farm out our children's education (and data) to Big Tech. It goes like this: the iPad becomes the teacher. The teacher becomes the iPad manager. And the children? They become an iPerson, the product of Silicon Valley's largest educational experiment yet (Herold 2019).

I was curious what personalized learning looks like in today's classrooms, so I asked veteran teachers and *Screen Schooled* authors Matt Miles and Joe Clement. "Twenty-plus kids in a room, all on their laptops or tablets, headphones on, working silently on their own devices with almost no interaction with each other or their teacher. How is that personalized?" Matt tells me, offering an example that hit a bit too close to home.

His son was struggling with his assigned math homework. Over and over, Matt heard the iPad buzz: Wrong, try again. Wrong, try again. Wrong, try again. "He was in tears," Matt says. After Matt offered to help with the problem "What two numbers make six?" he watched his son try 3 + 3. Wrong, try again.

"Even as an adult with moderate math skills, I had to think about it," Matt tells me. "Three and three do make six. What the hell is wrong with this thing? I finally thought to have him try four and two, and boom, that was what this program was looking for. Now, was that personalized learning? Would [his teacher] simply say 'Wrong, try again' and keep repeating that until he was in tears? Can you imagine a teacher that bad? But these programs are being sold as 'personalized learning devices.' It's the biggest marketing scam I can imagine. What a misnomer."

"Matt's son's experience is all too common," his coauthor, Joe, chimes in. "And [that's why] the term *personalized learning* might be the most dangerous one that EdTech proponents use," he says. "The phrase is brilliant marketing, to be certain. However, what [personalized learning] is, most often, is a kid getting online and answering questions. True 'personalized learning' means I know my students in a genuine, human way. I know when they need to be pushed. I can see confusion on their faces. I know when they are bored and ready to move on. Relationships matter. Human people matter. When a caring adult reaches out and says, in word or deed, 'I've got you,' that carries more weight for a child than any algorithm. . . . That deepened human relationship will do far more for that child in the short and long term than a happy-face emoji and on-screen confetti explosion will when they get a question correct."

The students hate it. The parents hate it. The teachers hate it. So what gives? "Using the algorithms to teach (and collect data) is simply quicker and cheaper for school districts," Joe tells me. "They view it as more efficient."

But it's not. Joe tells me about his third-grade daughter, who is required to take periodic math assessments online. His daughter doesn't use a phone or computer at home, so she routinely clicks or types the wrong thing on questions she knows and it's marked as incorrect. "Now she's in her head about it," Joe says. "So not only is this 'personalized' program bad for kids, it's giving the district bad information. If they simply asked my daughter's excellent teacher for an assessment of all of the kids in her classroom, they would get the real story."

But when it comes to EdTech, the story that matters is the one that pays.

In 2016, Zuckerberg and his wife, Priscilla Chan, announced their plan to donate "99 percent of their Facebook shares—worth an estimated $45 billion—to a variety of causes, headlined by the development of software 'that understands how you learn best and where you need to focus'" (Herold 2016). "We don't know for certain that it's [personalized learning] going to work," he said. "All we can really hope to do is provide an initial boost and try to show that it could work as a model, and hopefully it gets its own tailwind that carries it towards mainstream adoption."

A Guiding Light

So if Big Tech's formula of personalized learning isn't working, what might? And how can we as parents, as families combat the effects?

To answer, let's peek at the philosophies of famed nineteenth-century British educator and philosopher Charlotte Mason. Heralded as an innovative reformer of traditional education, Mason implemented a series of radical principles for a child's rich and holistic education, ranging from nature studies to habit training to read-alouds at all ages. Although a portion of her recommendations called for children to experience four to six hours in the fresh air to play freely (Mason did not believe in homework for younger children), she was anything but lax. Her third graders read Rudyard Kipling, studied DaVinci, memorized selected works of William Blake. Her students could identify famous compositions from the likes of Gustav Mahler, Edvard Grieg, the strings of Edward Elgar. By eighth grade, they'd have been introduced to the bulk of Shakespeare's works—*in Latin.*

Mason, whom the BBC recently profiled as a "guiding light" in childhood education (Lee 2022), pioneered a new approach to the way children should be taught. But it's her primary principle that should give us all pause, particularly in light of today's EdTech environment. In a grand treatise of her life's work, Mason writes, simply, "Children are born persons."

Were Mason alive today, she'd rally against personalized learning in favor of something larger: *personhood* learning. "Our crying need today is less for a better method of education than for an adequate conception of children" (Carman 2021), she writes. "Our aim in Education is to give a Full Life. . . . Life should be all living, and not merely a tedious passing of time; not all doing or all feeling or all thinking . . . but, all living; that is to say, we should be in touch wherever we go, whatever we hear, whatever we see, with some manner of vital interest. The question is not, *How much does the youth know?* when he has finished his education but, *How much does he care?*"

Mason's philosophy is the opposite of personalized learning: Caring, rather than knowing. Interest, rather than input. Personhood learning does not rely on a series of linear thoughts dispensed along a predetermined path toward what a think tank has deemed important. Instead, personhood learning connects children—mind, body, and soul—to the world around them. The real one.

And It Swims!

When my daughter was four, she spotted a new-to-her creature at our neighborhood pond. Ruddy cheeked and gasping with delight, she tried her best to explain to me in her limited vocabulary what kind of animal she'd seen. "A slimy tail with spikes! And a furry body! A head like a beaver, or maybe a squirrel? And it swims, Mom! It swims!"

Her grandmother suggested we consult Google Images when no one could decipher the description. But then I stopped. If I wanted my daughter to care, not know, wouldn't there be a better approach?

We called a neighbor who lived across the pond, a science teacher who spent more time fishing than anyone we knew. He knows this pond well, I told my daughter. He'll probably have some good ideas on how we can find out what you saw.

And he did. "A muskrat!" he announced with an excitement that nearly matched our four-year-old explorer's. "I'll bet those spikes

OPT OUT

Before you look to Siri for answers, ask someone you know. Think SIRI: Someone I Really Idolize. Not only might they have the answer but you'll have sparked a new connection and growing relationship along the way.

you saw were reeds. I think she's having babies, because she's building a nest in my back yard. Want to come over with your parents and see it?"

Hours later, we'd received a crash course on muskrat habitats, pond ecology, and biodiversity—far more than the quick response Google Images might have offered us. But even more, I learned a greater lesson on education. I didn't need to teach my children what to learn. I wanted, instead, to teach them how to learn.

I wanted to introduce my daughter to wide ideas and wonder-filled moments. I wanted her to seek out stories, to listen well, to learn from others. I wanted her to describe things with passion, to delight in the discoveries of the world around her. I wanted her to search for solutions beyond Google. To develop theories, to test them, to fail, to try again. To approach the world with curiosity and wonder, with open hands, not typing thumbs.

I wanted her to know, yes.

But more, I wanted her to care.

I Don't Use Cell Phones

Years later, I connected with a kindred spirit in my quest for a low-tech and high-impact educational experience. I first met award-winning author, neurophysiologist, and educator Dr. Carla Hannaford on a busy sidewalk in downtown Dallas, where she had just come from a mile-long jaunt down Commerce Street. "I'd have told you I was running late," she laughed, "but I don't use cell phones."

Carla is a vibrant soul with sparkling eyes who believes deeply in—and has dedicated her life's work to—the brain-and-body connection. "We take in our world through our senses, through our bodies, through movement. And we are really suddenly realizing that the top neuroscientists in the world are saying that the only way that we actually learn is through hands-on sensory input.

[Our hands] are the last organ in the body to fully develop. They're our greatest tool, and it's so vital that we do hands-on things. Swiping up on our devices doesn't work. It doesn't do it. We need to be out in nature, moving."

I asked what that means for parents of students who are learning on iPads throughout the day. How can those families support their kids once they're home? How can they counter that sort of digital input? "Any movement that integrates both hemispheres, especially the frontal lobes, where executive reasoning takes place and where the high level emotions reside—love, altruism, empathy, compassion, etc.," she says to me. "Integrated, cross-lateral movements do this beautifully—taking a walk, doing any of my Brain Gym movements, yoga, tai chi, etc. Even just physically playing with each other—like a soccer game."

+FOLLOW

To learn a few of Carla Hannaford's science-based, whole-body Brain Gym movements to help detox from screen overuse, visit optoutfamily.com/more.

Her methods don't just work for kids. "I really depend on [these movements] to keep me present myself" (Hannaford 2005), Carla admits.

Carla's groundbreaking work in whole body education is leading the charge for new educational movements all over the country. Hybrid schools, homeschooling co-ops, forest schools, outdoor education camps, and deschooling programs are growing in popularity for families seeking to return to a more holistic—and more effective—approach to their children's education.

Funnily enough, these programs all have one key thing in common: they're modeled after how school once was, but no longer is. "When I began my teaching career in 1969, open classrooms were all the rage," writes Dr. Jean Lomino, director and lead trainer of the Forest School Teacher Institute. "A few years later, walls went back up and open classrooms were a thing of the past. But I kept right on teaching the best way I knew—experientially, with as many field trips and nature-based lessons as possible."

Dr. Lomino is a champion in personhood learning, starting the first forest kindergarten in the state of Tennessee. She has taught countless students, trained teachers from all corners of the world, and presented at the International Symposium of Forest Kindergartens in Seoul. But one of her greatest accomplishments is that of her last years in the classroom with seventh and eighth graders. "My co-teacher and I focused our efforts on developing an environmental-education project. We began an in-depth study of the creek that ran through the community, doing water-quality testing, clearing trash, and advocating for our creek's health. The students eagerly got to work, establishing a legal 501c3 called Mission Environment, electing their own officers, and running it as a business. The result was a greenway built by the city along the creek. This was [personhood] learning at its best—and the parents and teachers were thrilled to see children pursuing learning with joy and enthusiasm because it was real and they were serving their community and helping to preserve the environment."

As I listen to her story, I'm met with a pang of regret for the children who will never get to experience this sort of ownership of and contribution to the natural world around them. Just yesterday, I received an email from our local nature preserve celebrating a new initiative to replace their communitywide, all-ages forest hikes. "Now introducing VIRTUAL HIKES!" the subject line read. "Join naturalist Amanda on Facebook Live as she hikes a trail, discusses seasonal ecology, and identifies species along the way. Participate in the live chat or just enjoy the scenery of the woods . . . from the comfort of your chair."

A Different Model

And yet there is hope, and it can be found in the unlikeliest of sources: Mark Zuckerberg's high-school alma mater: Phillips Exeter

Academy (PEA). A peek into the New Hampshire boarding school, which has a teacher-to-student ratio of 1:5, reveals a different model entirely: images show students collaborating, books scattered on every surface, chalkboard scrawls exploring all manner of subjects. While EdTech pushes personalized learning, individual results, algorithmic paths, Exeter abandons every ounce of that approach by centering their students around a device that isn't rectangle but oval.

A table.

They call the pedagogy Harkness, established in 1930 by Edward Harkness, a "man who believed learning should be a democratic affair." The concept? Twelve students for every one teacher, sitting around an oval table, discussing. "It's about collaboration and respect, where every voice carries equal weight, even when you don't agree," the school's website touts. "It's where you explore ideas as a group, developing the courage to speak, the compassion to listen, and the empathy to understand. It's not about being right or wrong. It's a collaborative approach to problem solving and learning. We use it in every discipline and subject we teach at Exeter" (Phillips Exeter Academy, n.d.).

If PEA's educational strategy is good enough for Zuckerberg, it's good enough for the rest of us, right? It sounds refreshing. It sounds revolutionary. It sounds . . . pricey. The good news: you don't have to pay the annual $64,789 tuition to try the model for yourself.

Because the Harkness pedagogy sounds a lot like what many homes—in every nation, in every language—have already been practicing for centuries: family dinner.

While many families can't control what happens inside the walls of their children's classrooms, we can—and must—offer a reprieve within the walls of our homes. One family I spoke with hosts Debate Night each evening with their four kids. Over dinner, each child takes turns choosing an issue that they're curious about, offering what they know about the subject already. "It has been a really great way to kind of rewire some thought patterns or societal untruths that have rubbed off on them," the mother writes. "It's through these conversations that my older kids can learn how to think critically, to prepare an argument, to listen thoroughly and well. And the younger kids get to see that all modeled. While I can't ensure they're getting that sort of communication practice at school, I can make sure they get it here."

Family walks, shuffleboard tournaments, baking dessert. Creating space to have open conversations, distraction and interruption free, is perhaps our greatest tool to use against a growing EdTech movement that sees our children not as people but as numbers and data and scores.

After all, to be more engaging than the algorithm, we have to actually engage.

One teen I spoke with wished her parents would make tech use a dinnertime topic. "They ask me about school, but school feels like it's all tech anyway, so it would be nice to have an open door to talk about that instead," she says. "And I'd love to hear from my parents about their tech pitfalls, too! I think that would build more of a bridge so it's like we're actually in this together, and it would help me to understand that tech use impacts adults, too. Kind of . . . see what we're in for later down the road."

"I used to let my child decompress on his iPad after school because he was just so tired from the day," one father writes. "But then I realized he wasn't tired, he was overstimulated. And the overstimulation was just compounded. We shoot hoops outside now, and it's like I've got my kid back."

"I've had to fight harder to keep my evenings free," says one mother I spoke with. "I recognize that I get my child for only a few hours every night, and even though I'm tired from my day too, it's just not worth it to use that valuable time shuffling him back and forth to activities. Detoxing him digitally and connecting as a family

⏻ LOG OFF

Not sure where to start with creating a more communicative, open dinner-table experience? Try conversation cards like Table Talk or Table Topics, or check out a book of riddles, logic questions, or conversation starters from the library.

has become our new extracurricular. It's harder than I thought, but I'm already seeing amazing results in the way we interact together."

While changing the future of personalized, digitally enhanced learning seems like an uphill battle, it's worth the fight to advocate for a generation of opt-out kids. "Educational policy makers are always interested in hearing what the parents/tax payers/voters in their district have to say" (Chicago Review Press 2017), *Screen Schooled* authors Matt Miles and Joe Clement encourage. "Reach out to school boards, superintendents, and principals to let them know your stance on technology. Most schools (if not all) have some sort of opt-out policy for device use. Some require device use in school but allow you to opt out of bringing one home. If enough parents start opting kids out, or at least asking questions and presenting information they have learned, policy makers will have no choice but to take notice."

⊙ DM

"I've talked with my [fourth-grader]'s teacher about my son opting out of iPad-assigned homework, and she's been really flexible about it. She agrees, actually, that they get enough time on screens during the day. As long as my son isn't falling behind, and as long as he can meet the alternative requirements, she's welcomed the opportunity to collaborate with us on our values. So now we use our at-home time to be together, cooking dinner, talking about our day, addressing things that come up. I feel like such a burden has been lifted for us all."
—Carrie W.

Potato Sacks

One such parent is Juliet Starrett. After noticing that the kids in her daughter's class couldn't manage getting in and out of potato sacks for a field-day race, Juliet founded a nonprofit called Stand Up Kids to combat daily desk-sitting in the classroom (Wallace 2015). "Educators are really seeking alternatives to the current environment," she says. "With all the technology and the amount of time that kids are sitting, teachers often bear the brunt of that in terms of behavior issues [and] attention problems in school."

District teacher of the year Kevin Stinehart has seen firsthand the benefits of movement and exploration extending beyond a child's behavior. "When our students' recess time was doubled and recess was offered multiple times a day . . . our behavior issues plummeted, math and reading scores rose substantially, kids were happier, healthier, and more primed for education," he tells me. "With basic social and emotional needs met, kids do better in every arena. I would extend that to just humans in general—adults, too, need their social, emotional, and mental health needs met in order to achieve goals, reach their dreams, and have a meaningful life."

The research stands: dirt, water, and all manner of natural elements are the building blocks of a child's education. Tactile skills, sensory play—they're an essential introduction to the world we live in. "Even the best educational computer programs and games, devised with the help of the best educators, contain a tiny fraction of the outcomes of a single child equipped with a crayon and paper," writes author David Sax in *The Revenge of Analog* (Sax 2016, 181). "A child's limitless imagination can only do what the computer allows them to, and no more. The best toys, by contrast, are really 10 percent toy and 90 percent child: paint, cardboard, sand. The kid's brain does the heavy lifting, and in the process, it learns."

⊕ +FOLLOW

Nonprofit advocacy group Fairplay has created an impressive toolkit called Screens in Schools Action Kit, which empowers parents to speak out against EdTech's harmful initiatives. Filled with templates for letters to administrators and newspapers, handouts for PTA meetings, research handouts, petitions, and opt-out forms, the kit is an exhaustive resource to aid parents in the fight against EdTech. To download a toolkit, visit *optoutfamily.com/more*.

Can it really be that simple? Paint, cardboard, and sand? If so, why are we—as parents, as a society—plowing past early-childhood guidelines in a race to be the first to offer kindergarteners a keyboard? To find out, I asked psychiatrist, author, and screen-time expert Victoria Dunckley.

"There's a persistent worry [I see] from parents: *If I continue to restrict access to technology, will my child get left behind?* But supporting brain integration by being as screen free as possible means you'll be optimizing your child's learning ability," Dr. Dunckley explains. "A child who has great computer skills but poor frontal lobe functioning will have trouble advancing in anything, since good frontal lobe function is needed to 'get things done,' tolerate frustration, and develop a strong social network. The frontal lobe is where creativity, innovation, discipline, 'big picture' thinking, and grit are born and bred" (Dunckley 2016).

Dr. Dunckley is quick to point out the research proving that computer skills are often overemphasized in schools, often to the detriment of other types of learning. "Rhesus monkeys can easily learn how to use a touch screen or joystick to problem-solve on a computer, and dolphins and apes have been taught to use iPads. It's just not that difficult."

Even for older children preparing to enter the workforce, studies have yet to find any correlation between technology skills and better wages. So what gives? Overstimulation, dysregulation, disruption of the body clock, poor concentration, brain chemistry,

⏻ LOG OFF

Remember: learning doesn't end when your child leaves the classroom, nor when he or she leaves your home. If you're concerned that delaying technology will hinder your child's future success, take a few moments to reflect on the many things you've discovered later in life. Just as you can learn to water-ski at thirty, knit at forty, and fiddle at fifty, your child can—and should—live a life of nonstop learning. So what if they learn a new passion later in life? Didn't you?

blood flow, hormone balance, and chronic stress levels? Is it really all worth it? Dr. Dunckley isn't convinced.

"Who will be left behind?" she quips. "The child who cannot concentrate."

✏ SELFIE

Whatever your child's schooling scenario looks like today, how can you infuse that experience with self-guided discovery, exploration, and curiosity? In what specific ways can you allow your child to reclaim his or her own education? This week, ask your child what he or she is most interested in knowing about the physical world. Ask, "What do you want to see or try or figure out? What feels confusing? What skills do you want to improve?" There are no wrong answers. Together, set aside an afternoon to facilitate your child's exploration of that interest or topic or skill. Resist the temptation to dive in with your solutions, and instead ask, "What do you think?" By allowing your child to get comfortable with his or her questions, you're letting your child flex the muscles of problem-solving, creativity, and self-education.

Chapter 6

Are You There, Siri?

Reclaiming Wonder for a Worn Generation

The more clearly we can focus our attention on
the wonders and realities of the universe about us,
the less taste we shall have for destruction.
—Rachel Carson

For what will it profit a man if he gains the
whole world and forfeits his soul?
—Matthew 16:26

TECH'S PLAYBOOK

Everything is knowable.

OUR PLAYBOOK

Life is a mystery.

I pull sourdough bread from the oven, heat the kettle. My kids busy themselves with their designated tasks: filling pitchers with water, sweeping floors, chopping fruit and cheese. My son sets the table, gathers forks, and I remind him we'll have guests for the afternoon—three girls I'm mentoring for the summer before they leave for college, pursue their paths.

But only two arrive.

"Where's Claire?" I ask.

"She's staying home," says one of the girls. "You know, because of the flash floods? We got an update from Alexa?"

Together, we look up. There isn't a cloud in the sky.

The Streets We Walk

In Silicon Valley, there's a shorthand for the science of reasoning: if this, then that (IFTTT). Every trigger has an action. Every action has a consequence. Every consequence offers new information, new findings, new data, and the path plays out until a pattern can be determined. Every here has a there, everything in its place, nice and tidy and figureoutable. *Certain.*

But as much as technology would have us believe otherwise, there is no algorithm that will shield us from the full human experience. No convenience will aid in the accurate prediction— and guaranteed safety—of life's storms, proverbial or otherwise. No formula for heartbreak, no pattern to suffering. There is little science baked into the reasons why a darkened cathedral, candle-light, and a booming chorus of robed sopranos singing hallelujah can bring us all to our knees. There is no reason why the sight of a giggling, diaper-clad toddler chasing backyard bubbles should conjure unspeakable joy.

Because the real world—this messy, churning, glorious world— relies not only on logical pathways but also on pathos: experiences, emotions, doubts, fears, moods in all of their complexity. The streets we walk, the people we carry, the tears we dry—they invite us to engage every part of the brain we've been given: memory, feeling, movement, balance, breath, *heart.*

That's what makes our real world just that: real.

Our world calls for surprise, for the unexpected. This spinning globe beckons us to be caught off guard, to be gobsmacked by mystery, to be utterly enchanted by a beautiful, cloudless, sunny afternoon in the middle of Alexa's ceaseless storm warnings.

Our world calls for wonder.

Tracking Cicada Song

When children are small, perhaps not yet tall enough to reach the steak knives but definitely tall enough to eat the dog food, their questions abound: *Why are you named Mom? Why is cheese yellow? Why aren't we going swimming? Why are you looking at me funny? Who taught the cicadas how to sing?*

We entertain their whims, try our best to withstand the endless chatter. But soon we needn't try. The world grows loud and our budding Copernicuses grow quiet all on their own.

"It's not that children are little scientists, it's that scientists are big children," argue Alison Gopnik, Andrew Meltzoff, and Patricia Kuhl in *The Scientist in the Crib.* "Scientists actually are the few people who as adults get to have this protected time when they can just explore, play, figure out what the world is like" (Gopnik, Meltzoff, and Kuhl 2001, 9).

Some go on to make great strides for all of humanity—in disease eradication, habitat conservation, aerospace engineering. Some toil with the ethics of their fields, such as resurrection biology or stem cell research. Some invent apps for the rest of us.

Still more simply stop. They grow up or grow out, and they can no longer be bothered to chart astronomical patterns and barometric pressures and the shrill songs of cicadas.

After all, technology can do it better, faster, more accurately. Why spend a life tracking cicada song? It's already on Spotify.

But science is not technology, and technology is not science. When we assume either are infallible or undeniable or omniscient, we abandon our own neural pathways for an algorithm's labyrinth. We dig ourselves into perspectives we no longer question. We limit our world only to what is certain, what is clear, what is measurable.

Which means, of course, we limit it very much.

In the Absence of Certainty

"I was writing a paper for my English class, and the assignment was to make sure we weren't using any fake news," one teen tells me. "So I started researching nutrition, like keto and stuff. But one expert would say something that was totally contradictory to another expert. Like, 'Cottage cheese is great because of this!' 'No, cottage cheese is terrible because of this and this and this.' And these were both reputable doctors! Both had really strong research for why they were right! It was impossible to figure out if either was spreading misinformation."

We've heard the term *misinformation* circulating in Supreme Court hearings, debated around dinner tables, shared via headlines, presented in school curricula, debunked in digital-literacy training manuals. We're trained to spot it, and Meta's trained to take it offline, fast. But what exactly is Meta removing?

"We define misinformation as content with a claim that is determined to be false by an authoritative third party" (Meta, n.d.c).

It's an astonishingly simple definition, one that—if taken verbatim (which is what AI is trained in: verbatim)—could render even the Bible as fake news. (After all, Chinese government leaders are currently rewriting Scripture to establish a "correct understanding" of the text [Kuo 2019], and who is President Xi

Jinping if not an authoritative third party?) Not only is Meta's definition incomplete, it's also illogical. Today's misinformation problem—and Meta's strategy for solving it—is rooted in a common logical fallacy: faulty appeal to authority, where Meta is using the "opinion of an authority as evidence to support an argument" (Wikipedia, n.d.c).

Is cottage cheese good or bad? It depends on who Meta asks and which answer it likes best.

So why does it matter? Because accepting an authority's opinion as fact means we stop thinking for ourselves.

We assume the experts know better. We don't pause to explore other perspectives. We don't entertain our doubts, our independent histories, our unique and specialized knowledge. We overlook hidden agendas. We seek comfort in consensus, in certainty. We no longer answer our own questions. Over time, we no longer ask them at all.

We stop listening to our gut; we have Siri. We stop wrestling with uncertainty; we have Google. We stop seeking counsel from our elders; we have TikTok.

Soon the truth we're searching for won't be ours. It will be manufactured in Silicon Valley. We won't have earned it from hours struggling with doubt, through hot tears or listening ears or long runs in the park. We won't have cried over it, Joni Mitchell's *Blue* on repeat. We won't have taken a walk, decompressed, sought counsel. We won't have journaled through our questions—waiting, wrestling, waffling—finding the answer arriving just when we need it most.

Why would we when there's a shortcut and it has proven faithful, and it's right here, buzzing in our back pockets?

Feeling lonely? Try Snapchat.

Feeling confused? Here's ChatGPT4.

Feeling depressed? Try BetterHelp.

Feeling sad? Here's Happify.

This is the world our children live in, where questions and

answers are processed in extremes. There is little need for gray areas when the solution is just—snap! ping! alert!—right at your fingertips. Why wait to consult your brain at all?

Our digital platforms weren't built to accommodate wider problem-solving, nor were they built to communicate deeper solutions. Character limits cut off all nuance, explanation. Memes are quick to crop out context. From bold headlines to quippy slogans to protest chants, online communication lives by the golden rule, "Whoever shouts loudest is heard." As a college professor friend of mine says, "If it doesn't sound good on an activist sign, these kids aren't sayin' it."

But when we seek—and gain—certainty in extremes, we get a society that lacks empathy, that fails to envision positive outcomes, that struggles to imagine common ground. We feel confused. Hopeless. Worn.

And then we hop back online.

Uncertainty of Payout

Silicon Valley knows, of course, the addictive elements these apps contain. The pull-to-refresh feature built into today's top social-media apps is designed to mimic a slot machine, guiding our brains to seek surprise.

It's called uncertainty of payout. "When someone likes an Instagram post, or any content that you share, it's a little bit like taking a drug," *Irresistible* author Adam Alter writes (Yates 2017). "As far as your brain is concerned, it's a very similar experience. . . . It's the unpredictability of that process that makes it so addictive. If you knew that every time you posted something you'd get a hundred likes, it would become boring really fast."

In our search for certainty, we find ourselves uncertain again. So once we do get a like? It feels good. Our brains release a bit of dopamine, and a bit more, and soon enough, we're on an intermittent drip of the same drug that—in high doses—causes anxiety, difficulty sleeping, mania, and stress.

Even if we're not posting to social media, dopamine is released as we consume it. Every funny skateboarding fail you see on YouTube, every joke you hear on TikTok, every heartwarming caption you read on Instagram—they all elicit a feel-good reaction. So you keep going. You look for another, and another. There's a reason the tech world calls page views "hits."

More novelty. More excitement. More fun.

But as the game plays on, what are we losing?

With every scroll of their thumbs, what will our kids see? What will they learn? What will they try? What will scare them, surprise them, shock them? What will they grow desensitized to? What will make them laugh? Cry? What will confuse them? What will they be ashamed of? What will they hide? What will they share? What will they gain?

What will they lose?

That, it seems, is the real uncertainty of payout.

Lights Up like a Christmas Tree

What, then, does it look like to deploy uncertainty of payout in our own homes—in a healthier way? What new, uncharted territory can we offer to our digital natives?

The answers lie, again, within social media's playbook. The way in is the way out.

"Surprise is still probably the most powerful tool of all," writes lead online strategist Scott Redick for *Harvard Business Review* (Redick 2013). And the research backs it up. Scientists at Emory and Baylor measured changes in human brain activity in response to a sequence of pleasurable stimuli, either predictable or unpredictable. Contrary to the researchers' expectations, the reward pathways in the brain responded most strongly to the unpredictable sequence.

"The region lights up like a Christmas tree on the MRI," says Dr. Read Montague, an associate professor of neuroscience at Baylor. "People are designed to crave the unexpected" (Sommerfeld, n.d.).

Julie Bogart, author of *Raising Critical Thinkers*, agrees. "I like to say that the least taught literary element in writing is surprise," Julie tells me when I press her for her own secrets of engagement. "Surprising language, plot twists, unexpected facts . . . everything depends on it. You will stop reading if there is no surprise coming. If you think you know what's ahead, you won't finish the book or the page or the article. So take that principle—the element of surprise, the element of subversion, of mystery, of risk, and adventure—and right now, apply it to your home. Is your home surprising? Is it in any way an adventure? Make it smaller. Is there anything to look forward to, today, in my home?"

Her idea holds water. Aliza, age seventeen, says she finds it "easiest to turn off my screens when there's a more fulfilling alternative than using tech right in front of me." Her advice for parents? "If possible, spearhead family activities (a walk, a game night, a mini road trip, a bake-off, etc.) in which everyone is so engaged and caught up in fun there's not even a want to pull out individual devices" (Bernhardt-Lanier and Kopans, n.d.).

Ever welcomed your ten-year-old home from school with a can of silly string? Left a love note on your teen's mirror? Served pizza for breakfast? Taken a last-minute road trip to see a giant bubblegum wall? Skipped work and school so everyone can play board games and stay in their pajamas? Surprised your teen with concert tickets? Built a rock climbing wall in your son's bedroom? Memorized the lyrics to your daughter's favorite Taylor Swift song? Woke up your son at midnight for hot fudge sundaes? Chased lightning bugs with your toddler? Made s'mores in the oven and camped out in the living room? Toured your hometown?

⏻ **LOG OFF**

Brainstorm a few ideas to surprise and delight your kids. Make a list and choose your top three. Then scatter them into next month's calendar. Reflect on what happens next.

Infusing surprise into your home, while exponentially worth it, can seem like a full-time job. There are expenses involved—both time and money—and on many days, we're all just trying to stay afloat. The good news: you don't have to turn your house into a 24-7 fun zone. In fact, you shouldn't. It's a surprise, remember? Keep things unexpected, sporadic. Less is more.

You don't have to bear the brunt of the work, either. "Give your kids the chance to surprise *you*," Julie encourages. "You don't have to create surprises for them nearly as much as you need to be open to the element of surprise in them. The next time [your child] asks you to look at what they are reading, doing, seeing, *stop*. Read, watch, be interested. Get inside that amazing mind of your child. It is all mystery and surprise in there!"

The Darkness of an Ozark Mountain

The beauty, of course, is that surprise is merely a foothold to something even greater, something that holds the power to help our children rise above a culture that's both obsessed with certainty and yet more uncertain than ever: wonder.

What makes wonder so pure and powerful is that it's an emotion technology can't yet—or perhaps ever—replicate. When we see a family of deer playing in a wild-flower meadow on Instagram, we are delighted, yes. We are entertained. We are moved by beauty and joy. But are we awed? Are we outside of ourselves? Beyond ourselves? Awakened with the recognition that our world is both massive and miniscule, spinning us into a shared connection with one another?

Are we amid wonder?

And yet when we are driving in the darkness of an Ozark mountain, a whole carful of us grumbling, impatient, tired, and there are no more dried mangos and the middle child has to pee, and we have just come round the bend in search of a gas station but instead we are met with the pricked ears and startled eyes of a fawn and her mother in the fog—*Just over there, right there, to the left, no, to that side right there, I am pointing at them, can't you see?*—and you slow the car to a stop, and no one moves, breathes, blinks until the twin enchantresses frolic away with the same awe of you as you feel of them? Now that's wonder.

Wonder is accessible to us all. It is free and available, ready and welcome in the vast planet that carries us forward. But wonder isn't something we *give* to our kids. It's something we give *back* to them.

Wonder is so easily lost in the rush of a pace, the noise of a mind. It is robbed when we attempt to make everything knowable, understandable, tidy. It is stolen, daily, in ways small and large: When we force practicing scales on a budding pianist. When we stargaze at Hollywood before Haedus. When we forbid our young ones to pet the geese, climb the aspen, chase the wind.

⏻ LOG OFF

Want to spark wonder in your child? Give them a tree—a wily willow in the park, a towering maple near your front stoop. Have them name it. Let them visit it daily, check on it, care for it. Knock on the trunk, see who lives inside. Paint a picture of it. Paint a picture in it, on it, over it. Hide books between craggy limbs, whisper secret codes to enter in. Let them leap and linger. Spit mandarin seeds from the branches. Hang a bee hotel. Lean a welcome sign. Let them climb. Let them fall. Mostly, let them be.

Wonder offers immense lessons in nuance, in observation, in seeing life as it is, not as we wish it would be. The fire that offers warmth to a few can mean destruction for miles. The sea that buoys our souls also buries its secrets. Forests, marshlands, deserts, slopes—might they bring shelter, freedom, adventure, entrapment?

Wonder invites curiosity. It is not the concrete lexicon of Silicon Valley's IFTTT. It is not "If this, then that." It is "Why this? Why that? What next?" It is discovery. It is exploration. Meandering. Mess. Observations and fancies and wild, awestruck imaginations. It is watching an ancient tree fall at the hands of a fresh lightning storm, wondering, *Why this one, at this time, and not that one, at some other?* Wonder might swap theories, trade facts. But it knows the secret: there is a sacred logic largely unknowable to us.

Yes, we can stay informed by the knowledge we seek. But can we also stay *unformed* by it? Can we allow ourselves to exist beyond it? Can we factor in our own observations? Can we walk around as humans on this banged-up planet, learning as we go? Can we reject the idea that everything, today, right now, is knowable, searchable, findable? Can we allow room for unexpected results, unexplainable miracles? Can we—gasp!—allow for future discovery along the way?

For ourselves, for our children, for our world?

Maybe an Elk

A few weeks ago, Ken and I enjoyed a short getaway to visit family friends who are constructing a lake lodge on the side of a Montana mountain. Between drives past glaciers and walks in the forest, I found myself marveling at the open sky, the wild flowers unknown to me, the cerulean of the water we waded. I asked a million questions of the various locals we met throughout our stay. Is this hemlock? Why are the chipmunks so small here? Who left these tracks? What does an emu sound like?

Everyone freely shared knowledge gained from field guides, conversations, years of collective experiences. But it was our friends' son—a buoyant and contemplative boy just four years old—who offered the wisest answer of all.

It's the first night of our visit. We stay up past our bedtimes and drive ATVs up, down, and around the mountain to see the newly framed walls of the latest construction projects. Pulling over at one final building, the husbands lose themselves in a conversation about trusses while, growing bored, their son and I lie flat-backed on a slab of plywood to count the stars above.

As the galaxies wink in the midnight sky, the surrounding animals arise to greet us with songs. We listen, quietly, eyes wide. "Do you think that sounded like an owl?" I ask my friend's son. "A raccoon? Maybe an elk?"

We wait. We listen. The animal call returns, again and again, echoing through the forests, sweeping through the breezy miles. "What do you think that is?"

He turns to me with a grin, whispering, "I think that those are things I'd like to think about together."

And so we do.

✏ SELFIE

Ready to offer your child a crash course in wonder? The four elements of water, fire, earth, and air were first recognized by the Greeks to be the four "roots" of which the world is composed. By engaging with each, you offer your child a firm foundation in the mysteries of our planet. Divide a sheet of paper into four columns: Water, Fire, Earth, and Air. Under each, list a few ways you can incorporate each element into the rhythms of your home. Can you wash windows with sudsy water, take a bubble bath, drink from the garden hose? Can you dine by candlelight, make shadow art on the walls, host a bonfire? Build a rock sculpture, make mudcakes, go barefoot? Fly a kite, blow dandelions, play catch? Brainstorm your own list, then circle your favorites to try this week.

Chapter 7

Game On

How to Challenge Our Kids and Encourage Healthy Risks

Difficulties strengthen the mind, as labor does the body.
—Lucius Annaeus Seneca, 5 BC–65 AD

They will soar on wings like eagles; they will run and not grow weary, they will walk and not be faint.
—Isaiah 40:31 NIV

❌ TECH'S PLAYBOOK

> ⬜ Online, you will be kept safe and shielded from real challenges.

✅ OUR PLAYBOOK

> ⬛ Offline, you will be made strong enough to handle life's challenges.

A violent snowstorm has pummeled our midwestern town, leaving downed power lines and fallen oaks. Tonight, my husband and ten-year-old daughter venture off to the neighborhood forest to clean up the damage, to bring home dead limbs for firewood. As the sky grows black, the two younger kids and I clean up the dinnertime dishes, read stories, take baths, head to bed. After the last of the lullabies, I am preheating the oven for cookies when my daughter bursts in: "Mom! We found a fallen black walnut tree! We'll be right back!"

I smile as I watch the back door slide closed, knowing better than to stifle an adventure. Hours later, I hear the two booted explorers finally clomp in, bone tired with rosy cheeks and buoyant spirits. "We did it!" they whisper. "Come see!"

I follow them out to the garage where a massive black walnut log sits in place of my car. I am astonished. It is, easily, 250 pounds. "How did you two get it here?"

"We used the little red wagon," my husband says, laughing with a shrug.

"And a whole lotta grit," my daughter says.

They scramble inside for cookies. And as mittens dry on the floorboards, as coats roughhouse in the dryer, as meltwater from their snow-caked hair drips onto fleeced shoulders, they dream up their next project: a handmade coffee table from their salvaged tree.

Tok Bottom

When it comes to social media—and particularly TikTok—challenges are all the rage. Baking a chicken in Nyquil. Showing your junk drawers. Covering a mirror in Vaseline and setting it on fire. Performing an Adele concert with gummy bears. Pretending to faint. Pretending to choke. Pretending to have died in the Holocaust (BBC 2020).

As a *New York Post* features reporter writes, "Humanity has hit Tok bottom" (Cost et al. 2023).

But the appeal of these challenges is understandable given everything we know of brain development in children and teens. "This desire to belong, paired with another key feature of adolescent brain development—risk-taking—is what leads to challenges . . . becoming such big trends," Teodora Pavkovic, digital wellness expert and psychologist, explains (Sager, n.d.). "Young people find a lot of meaning in getting involved in these trends, and taking part in them gives them an exciting avenue to explore what they love doing best: bending (and breaking) the limits of what adults expect from them."

Science agrees. Have you heard of hormesis, dopamine's antidote? In his acclaimed book *Antifragile*, author Nassim Nicholas Taleb defines the term this way: "A bit of a harmful substance, or stressor, in the right dose or with the right intensity, stimulates the organism and makes it better, stronger, healthier, and prepared for a stronger dose the next exposure" (Taleb 2014, 429).

Challenges make us stronger. And stronger. And stronger.

We know this, in theory. It's the reason we download the Duolingo app to brush up on French and why we keep kettlebells in the garage (whether or not we actually use them). But for children who are still in the thick of their brains' developmental stages, there's a key factor to consider. Not all challenges are created equally, nor are all children.

It's called developmental risk sensitivity theory. The theory proposes that "developing organisms [children] learn to use different risk strategies based on the availability of resources and the extent of their needs. A well-fed fox, for example, is unlikely to risk entering dangerous territory for a large meal when a small, certain amount of food is readily available," writes Peter Blake, a lead researcher and CAS associate professor of psychology at Boston University (Steinbrenner 2022). "A hungry fox, however, is more likely to take chances for a big dinner."

His theory offers profound insight into the false comfort TikTok's risky challenges offer our children. The promise, on TikTok, is that you'll be fed. You'll be found likeable and witty and charming. You'll go viral. You'll find friends and connection. You'll be satisfied, satiated, full.

It's worth the risk, right?

Not exactly. A year-long study into social-media culture by future-media researcher Debbie Gordon discovered that, increasingly, teenagers "are going to unsettling lengths to gain online fame." Gordon's research team found "thousands of fame-hungry teens, some with millions of followers, posting provocative photos or videos." Abandoning lessons they've been taught about online privacy, these children are publicly sharing near-nude content and "opening themselves up to exploitation and bullying."

"When sharing becomes over-sharing and that becomes fame-craving and that becomes obsessive, then obviously we need to start asking why," Gordon says (Clarke 2014).

But we already know why.

The foxes are hungry.

Cross-Dressing at the Smithsonian

Many parents I spoke with lamented that their child assumes it's easy to get famous on social media. But why wouldn't they? Every time they open an app, their pictures are displayed in the same feed as peer influencers, celebrities, and all manner of Kardashians. Online, the assurance of fame is just one click away. It's no mystery, it's marketing.

I asked a friend who runs an influencer PR agency to tell me what TikTok's algorithm suggests for young users who want to be noticed. She followed up with a link to TikTok's creator brief (TikTok, n.d.), a document designed to educate users on how they can better grow their audience, promote their work, and get TikTok Famous.

I open the brief to the first page, where I'm met with this promise: "You can be a part of any community you like! Just ask Creator *OnlineKyne*." The promo video autoplays and I meet Kyne, who self-identifies as TikTok's math-teaching drag queen. "[My] videos went viral and my community found me. I think the magic was due to the TikTok algorithm. You're getting to learn about new communities and subcultures which you never would even think existed or want to type into a search bar. . . . I mean, where else are you going to find a Filipino-Canadian teaching math while cross-dressing at the Smithsonian museum other than TikTok?"

Kyne's story is one our kids see often. Overnight successes, instant influencers, skyrocketing followers, brand deals, money, fame, attention, love. It's a story the algorithm scrambles to highlight, marketing one creator's success as something easily—and quickly—gained. But we know the truth. Overnight success is a myth, and attention gained is not always for gain.

For every Kyne, there are a thousand more Kates.

Kate joined TikTok when she was sixteen as a way to have fun, blow off steam, and join her friends. She loved to dance, loved music, and loved meeting new people. "I was hesitant to post anything, at first," she tells me over chai. "But I watched all these other people's dance challenges and it looked so easy and fun and effortless, so I thought why not?"

Her first post was liked by most of her friends. Her second one received even more comments. "Some were a little weird, and from people I didn't know, but I just kind of ignored those." But by the third post, the comments took a turn. "I'd call it harassment for sure," she said. "It was just terribly cruel, the things these strangers were saying to me and about me."

Many teens I spoke with showed me a string of what they call "mean-fairy comments," a TikTok trend where people leave insulting remarks disguised with hearts and happiness emojis:

I almost scrolled past this 🖤🗡 next time I make sure I will 🥀🗡

I'm pro-choice 🖤🏴 because of you 🔪🗡♀

The video is so nice! 🖤🌿 try it without you in it! 😊🗑

These mean-fairy comments range from mean, yes, to dark.

You ate this up ✨🗡 I recommend starving tho 🖤✨

You look tired 😵 go take a permanent nap ✨ u deserve it queen 🖤

Life but make it 🔪 end 🔪

She's jumping for joy 🎀🕸🦋🗡♂ I hope it's off a bridge

"I just sank," Kate said. "It's weird—you go on these apps talking yourself into posting, like, don't worry, the whole world isn't watching. Just have fun. But when you get something terrible in your comment section, it does feel like the whole world is watching. It really, really does."

Kate didn't show her parents the dances or the comments. She deleted her app the next day. "I just feel really lucky that it happened right away, that I didn't get sucked in too long," she says. "I think that would have been really hard to walk away if it was, like, my whole social life."

As she turns to throw her coffee cup in the trash bin, she laughs and says, "Just another failed experiment in the life of Kate!" But her eyes don't register the smile.

Get Messy

A few days ago, I found myself revisiting TikTok's brief and reading its advice for new creators looking for an audience on the app. Inside, I discover their tried-and-true recommended strategy:

- "The goal shouldn't be to 'perfect' a specific format or aesthetic. It should be to dive in whenever inspiration strikes.

- "Being passionate or sincere is more likely to get you on the [recommended] page . . . even if you're simple. Genuine. A little messy.
- "Fortune favors the bold. If you 'flop' (or think you flop) on TikTok, that means you tried something new. Which is what matters . . . you can try again. And again. And again."

"That's actually what I love about TikTok's challenges," says one teen I spoke with. "They're just for fun. Like, if I fail, no big. But in real life, if I fail, my parents are going to flip out."

As I pored over the brief, something clicked. Inadvertently (or not), TikTok had sung the ultimate siren song for a generation that felt risk averse, suffocated, and anxiety ridden. With the same airy nonchalance every tech startup seems to boast, TikTok flew onto the scene with the ready-made cure our high-strung, high-performing teens were seeking: permission to make mistakes.

Dive in. Get messy. Flop around a little! While TikTok's advice sounds harmless, what's the result for an entire generation encouraged to lower their inhibitions online for the sake of fun or fame? Harassment, for one. The worst possible outcome, for others.

Take TikTok's "blackout challenge." Challenge participants are encouraged to choke themselves with household items until they become unconscious, then film the adrenaline rush once they regain consciousness. But some never do. The trend has been linked to more than eighty deaths. At least fifteen were children twelve and under (CDC 2008).

One was named Mason Bogard.

"Our youngest son, Mason, didn't spend a lot of time on social media," Joann tells me. "He would rather be outdoors hiking or fishing or at his workbench creating something."

But he did watch how-to videos on YouTube, such as for

creating fishing lures and doing woodwork. After the launch of Shorts—a one-click feature that allows users to automatically upload TikTok content to YouTube—even kids without access to TikTok can stumble upon the app's trendiest challenges and top content creators anywhere in their YouTube feed.

"Young people view videos such as the choking game as safe and funny because no one is harmed in the videos they view," Joann tells me. "That is because the children who die never get to post their videos. For my son, the belt that Mason used locked in place and didn't loosen after he passed out. He never woke up. Mason was on life support for a week. We buried our sweet son's young body at fifteen years old. He had a whole life ahead of him, but he will never know that life."

Joann Bogard now spends most of her time advocating for parents against the online harms that face every home, every family, and every child. "We always said that Mason had an old soul," she tells me. "At a very young age he showed so much grace, mercy, wisdom and generosity. He shared anything that he had with others. He spoke up for anyone who didn't have a voice. So I try to follow [his] lead by speaking up for others. They won't always see the dangers online . . . until it's too late. Never think, *Not my child; they know better.* They are children. They aren't equipped yet to think ahead and know better."

We can download every parental control, install every data restriction, limit screen time, monitor use, block explicit content. But with Big Tech encouraging content creators to let loose, have fun, and continue to upload more than 150,000 new videos every minute (Hayes 2023), parents struggle to keep up with everything that's accessible on the screen.

"I don't let my kids have access to social-media apps, but even still, these challenges are the only thing they want to do when they get together with their friends," writes one mother. "At recess, after ball games—it's like they're always talking about a new TikTok challenge someone heard about. They're obsessed."

Gamification

So how do we compete with the constant lure of TikTok's challenges? How do we feed the hungry fox that craves acceptance and connection and distinction? How do we offer our opt-out kids a healthy risk—the safe, strengthening kind?

To find out, I looked to another maker of the Metaverse: UX gamification engineer and MediaSpark CEO Mathew Georghiou. A leading expert in simulation and gamification, Mathew's designs have reached millions of people around the world through schools, nonprofits, government agencies, and Fortune 500 companies. And for good reason. Gamification is the art of making things fun. For the same reason new parents fly a spoon as if it's an airplane to whoosh pureed peas into the mouths of their babies, the best gamification designers can transform any activity, chore, or task into an engaging, delightful experience.

"Gamification is really about influencing behavior," he tells me. "It's about motivation, extrinsic (external) and intrinsic (internal). Ultimately, intrinsic motivation has to be something inside of you, something that lights a fire in your belly, where your brain gets a little injection of dopamine that is keeping you interested and engaged. But also, good gamification has to have a proper flow, where it's keeping you connected and engaged so that just as you start losing engagement, for whatever reason, something happens that keeps you going. You feel like, oh, let me give it one more try. And you keep going."

It's what you—as a parent, and as a human being—do naturally. It's what we have been conditioned to do for thousands of years, swooping in with support or a listening ear or a helping hand, and not a moment too soon. Offering water to the parched. Offering counsel to the confused. Offering warmth to the cold.

Offering love to the lonely.

It's not the design or the content or the goal itself. It's *support* that's the gamification standby. "What I'm seeing," Mathew says, "is that the applications that use social networks so that you feel you have a support system, that you're going through something with other people, whether physical or online in some way? Those tend to be most effective."

Gamification, whether online or offline, relies on eight key elements to keep users engaged. Chances are, if you've ever played a board game, completed a challenge, or trained for a goal, you'll know exactly how to spot these elements—and how to incorporate them into your home.

Catch and Reward

The first gamification element, catch and reward, has immense power baked into it: positive reinforcement. The idea is simple: catch good or desirable behaviors and reward them. Our culture offers many catch-and-reward scenarios, for kids and adults alike: a craft store's loyalty program, your library's summer-reading contest, your Starbucks rewards. When you reward a user's behavior, you not only reinforce the continuation of that behavior but also raise the perceived value of that behavior.

"It's just human nature that people, and kids, too, want to be acknowledged and recognized, and they want to be appreciated. It's nice to be noticed," explains Judy Arnall, a Calgary-based author of four books on nonpunitive parenting, including *Parenting with Patience* and *Discipline without Distress* (Kadane 2022).

Sweden agrees. In a creative strategy pulled straight from a gamification designer's playbook, Stockholm law enforcement catches and rewards drivers who respect the speed limit. The motorists who are photographed driving slowly—catch!—are automatically entered into a cash-prize lottery—reward! "The people who are obeying the law will receive a portion of the money collected from those who were speeding," explains Kevin Richardson, designer of the innovative program (Haggarty 2010).

It's a far cry from the typical model of keeping roads safe: to punish bad drivers by use of fines, tickets, and, worse, license revocation. But does it work? Unequivocally yes. After the National Society for Road Safety implemented the program throughout Sweden, average speeds decreased from 32 km/h to 25 km/h (Howells, n.d.).

Just as a gaming app reinforces your child's constant use with enticing rewards like free lives, in-game trophies, or unlocking new levels of achievement, new companies like Aro are enlisting the power of gamification to encourage families to stay *offline*. Touted as the first app that pairs with a device to help you put down your phone, Aro helps families reduce screen time by rewarding users for less screen time and more intention. Families can set goals, get reminders, track time away, and compete with others—all in the time it takes to place their phones into Aro Home's charging station.

⊙ DM

"I recently disconnected my smartphone, but I really missed Wordle! So I introduced Scrabble to my granddaughter. We have popcorn and make a whole night out of it, and by the time the night is over, I know every one of her deepest secrets and boy crushes! It's my favorite part of the week." —Tammi S.

Whether or not you rely on an app to gamify your family's behavior, every parent can utilize this strategy to support desirable habits within their household. Do you thank your children when they speak kindly to each other? Do you take your kids out for ice cream when they stand up for what's right? Do you stop your child on a random Tuesday morning—amid the chaos of packing backpacks and making lunches and burning eggs—and look them in the eye, tell them how cool it is to watch them becoming the person they are?

Then you know firsthand the mounting value of catch and reward.

Spot Motivators

In my conversation with Mathew, he mentioned the importance of motivators in the gamification industry. Everyone's motivators are different, which makes gamification tricky. "As a designer, I can't always know exactly what will motivate that particular person at that particular time," he says.

But *we* can.

"As parents, we can apply these [gamification] methods in our daily lives just by knowing each other and what our triggers are," Mathew says. "I think of my son with special needs. I found early on that [he] responded to two things—stories and competition. So if [my son] is taking forever to put his shoes on, I'll make a race out of it. 'I bet I can put my shoes on before you can!' I'll say, and all of a sudden, he's engaged and into it. Turning things into a game is a big motivator, it's very rewarding, and, if it's fun—stories, games—it doesn't have to be stars on the refrigerator, you know? Just simple little contests and competitions."

One mother I know was tired of asking her kids to hang up their coats. Instead of nagging or giving up altogether, she decided to secretly leave a treat in the pocket of any coat that was hanging in its proper place. "My six-year-old came running up to me like it was Christmas morning because he found a jelly bean in his pocket," she tells me. "I winked and said, 'Good things come to a coat in the closet!' I haven't tripped over a parka on the floor since."

Another parent I spoke with hosts a Six-Minute Tidy Tournament when the house is in serious disarray. He throws on shoulder pads and a helmet, sometimes even swiping face paint under each eye. "I even have a whistle!" he grins. "We do different challenges and unlock different levels, like picking up stuff with your left arm for one minute, or trying to clean with the lights off for two minutes. The kids love it."

If you're unsure what your child's motivator is, think of how they like spending their birthdays. Do they want a big party with lots of friends? Maybe social connection is a motivator. Would they rather have lots of small trinkets to explore over time, or one big experience gift to open and enjoy right away? How much activity do they want planned? Obstacle courses? Whipped cream fights? Do decorations or themes matter?

Spotting motivators is as simple as paying attention to your child's preferences. And if kids can be relied on for anything, it's telling you exactly what they want.

Set Goals

Goal-setting is a key element in gamification strategy, and it's also why playing pretend is often hard for adults. What's the goal? What's my role? Where's the script? We don't know what's expected of us, which can make the experience feel pointless. When children experience challenges, they feel the same way. To push past the threshold of a challenge, it's imperative that they're motivated to see it through to the other side.

The best gamification design makes those goals clear and measurable. Think of *Minecraft*, where users must gather materials and build shelter, find food, or make a bed in order to make it to the next day. The goal? Survival. Or Monopoly, where players buy real-estate properties to collect more rent than their opponents. The goal? Wealth. Or *Wii Fit*, where users can complete challenges in a variety of yoga, strength-training, aerobics, and balance mini games. The goal? Health.

Setting goals is a way of reverse engineering the gaming process. By starting with a goal, players can decide which choices to make, which skills to build, or whether they even want to participate.

By setting goals together as a family, you're working together toward an active role in your shared experience. Start by asking your children what their goal is. It might be small and measurable—*I want to learn to play the drums!*—or large and unwieldy—*I want to be happy.* But by reverse engineering both, you can break down that goal into smaller micro goals or challenges to help offer family support along the way.

Not all goals will be manageable or advisable. But by keeping an open mind and ongoing conversation about present and future goals with your children—and the entire family—you'll eventually land on something you'll be thrilled to get behind together.

"Setting goals helped me get my kid back," wrote one mother. "My son and I had gotten to this weird place where I could barely recognize him anymore. He wasn't the boy I had raised. He was becoming this combined result of everything he'd been told online. It was like he was the internet's kid."

So she started a conversation over dinner about goals and asked him what he wanted to learn how to do in the next year. "Surprisingly, he wanted to start a garden," she said. "And we did. We logged so many hours in the dirt that summer. We had some of the best chats I'll never forget."

He's no longer the internet's kid. Now he's an opt-out kid.

⊙ DM

Remember: you're invited into this process too! One of my goals a few years ago was to become less reliant on search engines. I hunted down the most difficult crossword puzzle I could find and hung it on the fridge. Every time a friend or family member visited, I'd ask them to help me with whatever number I was stumped on. The result? I'm still only halfway through the puzzle, but I've learned that, collectively, my shared community houses all manner of wisdom, knowledge, and trivia that I wouldn't have discovered otherwise. (And I now know that history's earliest recorded contraceptive was, unsettlingly, crocodile dung.)

In our own opt-out family, we've set and completed many goals together throughout the years, from small (learning Mandarin, memorizing Kipling, designing our own toys, hosting neighborhood art sales) to large (relocating to a remote mountain, where, if all goes as planned, each of our children will be granted the opportunity to design and build their own tiny home as their senior year project).

Shared goals can often provide a bridge between our child's experience and our own. Even if the outcome isn't valued by everyone—Why build a tiny home?—the process of planning, the exertion of energy, the creativity of problem-solving, and the making of memories? Those shared moments stack up to a familywide bond where you're not only championing a cause but also championing each other.

And perhaps most important? Goal-setting places your child—and your family—squarely in the direction of their dreams. Rather than being co-opted by the plans of a multi-billion-dollar app that profits from your child's passivity, he or she is walking toward a true, unique, and self-guided passion that points to a more promising future.

Track Progress

Once your child has a goal in place and begins to walk toward its accomplishment, there will be inevitable hiccups. Unlike what TikTok promises, success doesn't happen overnight. To help keep your child from becoming discouraged and abandoning the journey, take a cue from Snapchat's Streaks feature.

"Gamification elements, such as Snapchat's 'streaks' feature, which publicly keeps track of how many days in a row you've used the app, make users feel compelled to check their apps every day in order to keep up their rating," reports Catherine Price for Science Focus (Price 2018).

Every day indeed. A group of teens I spoke with shared how hard it is to take breaks from Snapchat because they don't want to ruin their streaks. "I lost my phone once, so I asked my best friend if she could log in for me just to keep my streaks going," one laughed.

"And then there was that time I was in the ER and you snapped for me!" says another.

"Monitoring goal progress is a crucial process that comes into play between setting and attaining a goal, ensuring that the goals are translated into action," says author Benjamin Harkin of the University of Sheffield (American Psychological Association 2015). The better you are at tracking your progress, the more likely you are to achieve your goal.

It worked for Jerry Seinfeld. When a budding comedian approached him for joke-writing advice after a night club performance, Seinfeld encouraged the young man to hang up a big annual wall calendar—a whole year on one page—and get a red marker. "He said for each day that I do my task of writing, I get to put a big red X over that day. After a few days you'll have a chain. Just keep at it and the chain will grow longer every day. You'll like seeing that chain, especially when you get a few weeks under your belt. Your only job next is not to break the chain" (Trapani 2007).

Tracking progress—whether through calendars, spreadsheets, bullet journaling, or visual systems—is a proven technique to realize your goal. Whether your child's goal is to become a better babysitter, buy a horse, become a black belt, or climb Clingman's Dome, trackers are one of the best motivators for getting there.

Just ask Ginny Yurich, founder of 1000 Hours Outside. "This may come as a surprise to some people since we lead a global movement with an emphasis on getting into nature, but our family doesn't naturally want to go outside," she tells me. "But the reason

+FOLLOW

If you share the goal of getting your family outdoors, find 1000 Hours Outside time trackers on *optoutfamily.com/more*.

we have a goal in the first place is because getting outdoors isn't our first choice. Given the alternatives, we would gloss over it almost every time."

So she created time trackers for her family—and to offer free to everyone who shares the goal of more outdoor play. "Once we are out there under the blue (or usually gray) Midwest sky, once we've passed the threshold of leaving the temperature-controlled, bug-free environment, we almost always enjoy our time outside, but getting out the door can be a struggle."

As with most things, getting started is the trickiest part. But gamification has the answer for that, too.

Invite Community

What would TikTok, Snapchat, and Instagram be without the friends you follow on them? How fun would it be to play war alone? How about hide-and-seek? Checkers? Tag?

Gamification designers know the importance of community, shared experiences, and feedback. It's the reason today's most addictive video games, like *Fortnite*, *League of Legends*, *Minecraft*, *World of Warcraft*, all contain social elements like forums, chat boxes, and leader boards to foster a sense of belonging and togetherness. The power of peer influence is a key element that gamification designers are quick to utilize in the addictive technologies they craft.

"I've tried to quit gaming so many times, but nobody understands it's not just a game," one teen tells me. "[*World of Warcraft*] is my whole life. All my friends do it. If I quit, I won't have anybody to hang out with or talk to."

Another teen was quick to point out how his social circle was affected by not gaming. "My best friend started playing *Fortnite*, but my parents wouldn't let me, so I kind of drifted apart with him. He's got this whole group now and all they do is eat/sleep/game. I feel kind of lucky I never started playing. So many kids at school are totally sucked in, like, not showering or doing homework."

But the same gamification elements that can diminish a life can also grow one. Ginny, who founded 1000 Hours Outside, knows the importance of relying on gamification elements in her kids' routines, especially ones that spark community. "Friends are the biggest motivator for us to include more outdoor adventure in our lives," she says. "They also provide accountability. If we make plans to meet up at a park or for a hike, then we're going to show up. And when we do, these experiences provide rich feedback, a reminder that it's worth it to step out of the comfort and the control."

What friends can you rally to join you or cheer you on? What shared community can you gather to work toward a common goal? Who will you bring on your journey?

The beauty of inviting community is this: you don't have to go it alone. You needn't start from scratch. Perhaps your local community already has a system in place where you can walk toward your goal slowly and simply. Is there an organization that collects coats for homeless neighbors? A neighborhood project painting fire hydrants? A Save the Sea Turtles initiative to oppose single-use plastics?

Sign up, and watch what happens next.

Experiment Freely

One of the joys and inherent values of gamification is the ability to assess, learn, and mediate potential risks within a safe and secure environment. When TikTok advises users to "dive in and get a little messy," they're getting it half right. The freedom to experiment is a gift every child should have. What TikTok doesn't provide, however, is the safe and secure environment needed to mediate risk. But we can.

Risky play outdoors—tree scaling, rock climbing, paddle boarding—offers immense benefits from experimentation, self-regulation, and risk assessment. Nature's environment is ripe with learning opportunities—*Is this an edible plant? Is this a poisonous snake? Is this tide too strong? Is this log stable?*—to help our children assess risk. We need only let our parental fear of inconvenience and injury get out of the way. According to medical researchers at

British Columbia's research and advocacy group Outside Play, recent estimates show that children would have to spend about three hours per day playing every day for an entire decade before they would likely get an injury that needs medical treatment—and even then, it likely would be minor (Outside Play, n.d.).

"For younger adolescents, parent(s) might design the environment to provide safe activities that focus on their need for sensation-seeking," writes Nina S. Mounts (Magliano 2015). "For example, outdoor activities with peers, such as rock climbing and zip lining, can provide a great context for providing the excitement and social relationships needed for young teens."

Or better yet, let them find their own way. "Self-education through play and exploration requires enormous amounts of unscheduled time—time to do whatever one wants to do, without pressure, judgment, or intrusion from authority figures," writes evolutionary psychologist and research professor Peter Gray (Gray 2015, 100). "That time is needed to make friends, play with ideas and materials, experience and overcome boredom, learn from one's own mistakes, and develop passions."

As our kids grow from footed pajamas to soccer cleats, they need space to interact with a wide variety of people, new opportunities, fresh experiences. They need room to breathe, permission to try, license to experiment. If we don't give them that offline, they'll seek it online.

Kite flying. Hopscotch skipping. Stone stacking. Many scientists and child psychologists posit that the physical world—being in nature, in fresh air, and engaging in risky play—is the treatment

⏻ LOG OFF

When you're out running errands to a familiar place, grant your kids freedom to lead the way whenever possible. Can they pick out the limes for you? Can they run ahead and you'll meet them by the spices? Can they pop in and pick up the library holds for you? Let them walk in front of you as you follow a few steps behind. Treat them as fellow people, valued humans. That's what they are.

required to combat many of today's mental-health crises, such as depression, anxiety, and mood disorders.

The short of it is this: In nature, your child is likely not only to be safe but to be *saved*.

Fail Safely

In my gamification research, many of Silicon Valley's programmers pointed me to Four Freedoms of Play, a rule widely credited to MIT professor Scot Osterweil. The concept posits that to gamify successfully, apps must incorporate four freedoms as a means to keep the user both engaged and motivated.

"The first of Osterweil's freedoms, and possibly the most important of them, is the freedom to fail," says gamification manager Bjørn-Rune Hanssen. "Games by their very nature allow the player to fail without persistent and harmful consequences, as opposed to the real world where failure usually comes with repercussions. . . . [This tolerance of failure] allows you to try different strategies, use different information, and accomplish your goals in different ways. Maybe the way you did it first was the best. Or maybe you find a way that is fifty times quicker and simpler. . . . There is no cost to trying" (Hanssen 2017).

Encouraging our child's failure takes practice, but, fortunately, not a lot of time. "Just seventeen seconds," according to Mariana Brussoni, a professor at the University of British Columbia and BC Children's Hospital (Toole, n.d.). Brittany Toole writes, "Instead of telling your child not to climb so high or run so fast while observing them at play, take a moment—or, as Brussoni advises, seventeen seconds. Step back, [Brussoni] says, and 'see how your child is reacting to the situation so that you can actually get a better sense of what they're capable of when you're not getting in the way.'"

By pausing before responding, you're offering your child a higher tolerance of failure and a chance to develop an increased level of risk assessment. It's a key developmental tool that many leaders in childhood education—like AnjiPlay, an internationally acclaimed and publicly funded early childhood program—are beginning to adapt and integrate into their curriculum. A quick stroll through the schools that incorporate AnjiPlay's methods reveals an abundance of planks, ladders, barrels, mats, carts, and climbing cubes, likely none of which would pass today's Consumer Product Safety Commission standards. But research has found that children playing and learning in this type of environment "suffer fewer injuries, and less serious injuries, than their counterparts at more traditional schools." As Cheng Xueqin, the creator of the AnjiPlay approach, writes, "It isn't how high you climb that makes it safe or dangerous, but whether you climb as high as you choose" (Cohen 2023).

Why? Because when children can develop their own tolerance for failure, they're safer. "When adults step in too much," writes Lawrence Cohen, author of *Playful Parenting*, "then children don't have a chance to practice their capacity to assess risk for themselves. That's just as dangerous as walking away and leaving children unsupervised. As someone who spent a lot of time saying 'be careful, be careful' to children, this was a powerful lesson for me" (Cohen 2023).

⏻ LOG OFF

Not only can we offer our children the space to attempt challenges, but we can also offer them the home environment to regularly overcome them. Can your toddler carry a hefty sack of flour in from the grocery? Dust the baseboards? Climb onto a step stool to rinse the dishes as you load? Our children display astonishing capabilities once we allow them to.

One mother I spoke with has replaced her own "Be careful!" mantra with a new one: "Do you feel safe?" She says, "I want [my son] to learn to trust himself when he can, rather than relying only on my assessment of safety. Yesterday, he was climbing trees with his cousin, and she was asking him for help getting to a higher branch. 'I can't,' he said. 'Because if you can't get up safely, you can't get down safely.' I thought it was a really profound thought. It proved he's learning to identify and process risk."

Share Achievements

Nearly every gamification expert I spoke with mentioned the importance of sharing achievements. From badges to rewards to unlocking new levels of opportunity, celebrating milestones is integral to creating consistent engagement over time.

It's why your Peloton app has a share feature so everyone on Instagram can celebrate (envy?) that new personal record you just hit. It's why you proudly deck out your daughter's Girl Scout Brownie vest with badges, pins, and insignia for every milestone she hits. And it's why your son can't peel himself away from *League of Legends* until he sees his avatar on the live-ranked ladder.

My friend Ginny knows firsthand how essential the process of sharing has been in her family's growth and achievement of goals. "When I look back on my own life, I can distinctly point to the influence of others through the books they've written and parts of their lives they have shared," she tells me. "In our 1000 Hours Outside community, I'm sharing our lives. I'm sharing what we have found that works. I'm sharing the things I'm reading, or the people I'm talking with who continually remind me to value real living."

It works. Sharing her achievements enthusiastically has helped Ginny guide the way for more than a million kids to join her family in the quest to balance out real life with virtual life (Reilly 2023). "We're cheerleaders," she tells me. "We are building a community that celebrates the effort taken to savor a playful childhood and a playful life."

You're likely already a pro at celebrating your child's achievements and milestones—and you've got the tooth-fairy bill to prove it. But don't forget to allow this gamification element to spill into other families in your sphere. See families talking together at a restaurant—no phones in sight? Tell them you noticed and offer a high five. Witness a parent at the playground giving a child space to take a risk? Encourage her. Emboldened by another opt-out family you know? Give them all the credit in the world.

Your Ear Is Bright Red

Shortly after giving birth to my third child, I learned I have a rare autoimmune disease. In the midst of a flareup, I wake to a painful, inflamed ear that progresses into red, rashy swelling throughout that side of my face. Once the fever and hives begin, I know I have just a few hours to call the doctor for antibiotics before my eye swells shut.

Mostly, this chronic disease is manageable, albeit frustrating. As anyone with an autoimmune condition knows, the timing of the flares is a mystery—unpredictable, not easily traced. This makes things tricky when scheduling important family dates—weddings, camping trips, house projects. Mostly, we try to be flexible. We do our best and plan in pencil.

But one weekend, Ken is scheduled to leave town. He's packing his bag, folding shirts and wrapping the toothbrush charger when he spots it. "Your ear is bright red," he says.

I send him on his way and swallow my first round of antibiotics. But hour by hour, the rash progresses more quickly than normal. The pain hits. The fever. The hives. I call my ten-year-old into the bedroom.

"I'd like to hire you as a babysitter for the weekend. You'll earn $100. Are you available?"

She beams, nods emphatically, says that yes, yes, she can absolutely run the household without me. I walk through a few basic ground rules for her six-year-old brother, a few more for the almost two-year-old. And with that, she's off.

I shut the door to sleep off the fever. Hours later, I awaken to sweaty sheets and a darkened bedroom. I pad down the hallway, turn on a dimmed light. The kitchen is clean. The dishes are finished. The floor has been swept. There is soup in the pot. Everyone is peacefully, happily asleep.

After Ken arrives home and I'm recovered and rested and we all settle back into a rhythm, my daughter tells me of everything that went wrong. Smashed blueberries in the rug, backyard tantrums, a prolonged bedtime routine—*We couldn't find her stuffed piggie! It took forever! It was in the paint drawer!* We laugh about how crazy the dinner hour feels, about how, on some nights, you have to pull an Erma Bombeck and just "throw an onion in the oven."

The moon starts to rise, our tea turns cold. As she gets up to brush her teeth, I thank her again for her hard work.

"Oh! And by the way, your soup was amazing," I tell her. "Which recipe did you use?"

She tells me she didn't use one, "just threw something together with what we had." She shrugs, turns on her heels, walks away— three feet taller in a single weekend.

✏ SELFIE

Reflect on the past few years in the life of your child. What has he or she overcome? What milestones crossed? What challenges met and mastered? Write them all down. This week, take a few moments of undistracted time with each child and celebrate those achievements together. Fix their favorite beverages, and toast their growth, accomplishment, and grit.

Chapter 8

AutoPlay

Building Momentum, Creativity,
and Flow into Your Family Culture

You can discover more about a person in an
hour of play than in a year of conversation.
—Richard Lingard

Provide for me a man who can play
well and bring him to me.
—1 Samuel 16:17

❌ TECH'S PLAYBOOK

> 🔘 Our built-in technology can offer endless opportunities for fun.

✅ OUR PLAYBOOK

> 🔘 Our built-in creativity will offer endless opportunities for fun.

One morning, I wake to find my husband hasn't yet slept. He is sitting by the fireplace, cozy in a wingback chair, eyebrows furrowed with focus. He is reading, legs crossed, wearing house slippers, and he looks as if he belongs in a Fitzgerald novel, with one exception: in place of a pipe, he is holding an iPad.

"I'm still trying to figure out the parental controls on this thing," he tells me. Our six-year-old son has taken a new interest in piano, and there's a piano skills app we'd found years ago. Ken dug out his old iPad from the closet downstairs, but I can sense we both carry the same trepidation: *Is this a good idea?*

When we lived in Los Angeles, Ken took a job with TBWA\ Media Arts Lab, the "bespoke global agency for Apple and only Apple." Entering with a thumbprint, he joined meetings and conference calls where plans were hatched and decisions were made, amenably and unanimously, like a well-oiled machine. He speaks the language. He knows the product. He understands the technology, and also the people who make it.

"There's no way this isn't on purpose," he says in frustration as he shuts down the iPad, saving the research for another day.

Big Tech prides itself on the availability and accessibility of parental controls. It's the ultimate buffer from any and all liability: *Here's a step-by-step tutorial! Here, a how-to video! We're here to help keep your kids safe! Chat with us anytime! Let us walk you through the simple process!* And it is simple, or at least it appears to be. If you're merely trying to set up controls, that is. But if you're trying to set up controls that a child cannot quickly, covertly circumvent? If you're trying to set up controls that will prevent your child from being enticed to stay on the device just a little bit longer? Impossible.

It's a frustration we've all heard parents share throughout the years: the many ways in which children can so easily remove the tech guardrails we put in place. If you place limits on text messages, your kid can just ask Siri to send them for you. If you place restrictions on gaming apps, those restrictions disappear once your teen deletes and reinstalls them. If you activate screen-time limits, the iPad will prompt your child with a friendly message to "Ask for more time" and click "One More Minute!" (Xanadu87 2022), like a wily grandmother who, upon granting your kid just one more cookie, winks and says, *Shh . . . don't tell your parents! It'll be our little secret!* (We're just purchasing friction, remember?)

From recording private YouTube videos for a midnight playback to changing time zones to manipulate screen times, tech secrets and work-arounds are passed around the playground like kickballs and fist bumps. It doesn't seem worth it, does it?

But sometimes, it does.

Because you have learned to play the piano, and your husband has learned to play the piano, and your daughter has learned to play the piano, and your son would like to learn the piano too, and could he learn using the same app everyone else used?

Whatever Plays Next

Recently, I asked a thirteen-year-old what she does for fun. "I don't have a lot of free time," she says. "But when I do, I'm really too exhausted to do much. I just, like, get on YouTube and watch vlog-brothers for a couple hours."

I asked her what she does once vlogbrothers is over.

"Whatever plays next."

She's talking about autoplay. Most popularly deployed on YouTube and Netflix, the feature ensures videos will keep playing without any engagement from the viewer. Finished with one video? Another plays. Finished with *that* one? Here, we found this thing you'll like. And once that video's over, we have dozens—no, hundreds!—more to suggest for you. Don't worry, we'll keep 'em comin'! Need more screen time? Ask for another minute!

"Companies have systematically removed stopping cues—those brief moments, like reaching the bottom of a screen, that suggest you might want to move on to something else," Adam Alter,

New York University psychologist and author of *Irresistible*, tells BBC Science Focus (Price 2018). "Where Facebook, Twitter, and Instagram make their feeds bottomless, introducing natural endpoints would gently encourage users to move on to other activities."

But, of course, they don't. Because introducing natural endpoints might nudge us elsewhere, back into *real life*, which means less time on device for the platform, which means less profit, less metrics, less data, less attention, less addiction.

Unlike Netflix and YouTube, many social-media apps take autoplay a step farther by removing both stopping cues *and* starting cues. On TikTok, there is no play button. You simply open the app and, instantly, you're watching whatever TikTok is playing. The effect is not unlike entering a room where a social gathering is already happening, and TikTok knows the appeal. "It's Death Row Day and everybody's celebrating!" displays a recent TikTok promo. "So, roll down the street, sip the beverage of your choice, and let this incredible music inspire you to be as creative as you can!" (TikTok 2023).

One teen I spoke with describes TikTok as "the party I'm always invited to—and it never ends!"

And the band plays on. Pounding music, three-second attention grabbers, bold headlines, AI filters, flashing graphics. TikTok's "party on" platform boasts itself on encouraging creativity but steals the quiet contemplation and originality required to enter into creativity.

And once you're bored (if ever)? Switch to Snapchat. Or Instagram. Or YouTube. Just keep clicking, please, *one more minute.*

It is this—the starting cues, the passivity, the mindless lure—I am thinking of when my husband and I finally land on a solution for introducing an iPad. It will live at the family piano, resting upright in place of sheet music. It will be plugged in to the receiver, not made portable. It will be single use, with no other apps installed, nor internet. It will not be called a personal device or even a shared device. It will be, simply, an instrument for music.

This is how our opt-out kids have learned to play Canon in D, *Für Elise*, and, of course, "Baby Shark." And their toddler sister? She has grown up with the innocent assumption that every child on a tablet—in restaurants, on airplanes, in waiting rooms, at the grocery store—is simply practicing scales.

Apple's devices are a Swiss Army knife of stimulation, a design decision that prevents our neurological pathways from entering a deeper state of focus or relaxation. Much of today's tech addictions stem from sifting through the vast amount of opportunities for distraction that beckon us hourly. *Weather! No, Twitter! No, Wordle! Just one more minute!* We've all tried methods of reducing the constant notifications, pings, and buzzes, but as recent research suggests, they're ineffective at best. Why? Because the very sight of a smartphone—even when turned off—short-circuits our neural pathways.

Studying a phenomenon called brain drain, scientist Adrian Ward of the University of Texas led a team that conducted experiments to test how the mere presence of smartphones affects our ability to think and function offline. Their conclusions were shocking. "Placing a smartphone face-down or turning it off do not appear to lessen the brain power it consumes" (Trilling 2017). "The only thing [we] believe will help people who wish to increase their focus and cognition is to separate themselves from their smartphones—by, for example, placing the devices in another room."

Psychiatrist Anna Lembke, author of *Dopamine Nation*, is hyperaware of the detrimental effect of smartphones on cognition. "I don't own a cellphone," she tells *Stanford Daily* (Park 2018). "I do believe when we're constantly having our train of thought interrupted by checking a message or checking a text, we deprive

OPT OUT

Delay introducing iPads until there is a compelling, personal, and specific reason not to. By keeping a device single use only, you are reducing it to a piece of equipment, rather than a piece of entertainment. Treat an iPad as you would a waffle maker. If you're not making waffles, it's useless.

ourselves of having a sustained flow of thought, which is crucial to creating something. Yet by constantly checking and responding on the smartphone, we have the sensation of doing or making something. But it's an illusion, because at the end of the day, we haven't created anything. We've only been in response mode."

Only by inviting expansive quiet can we move out of response mode into the widening boulevard of self and the wilderness beyond. Room to breathe. Space to dream. Time to think, to reflect, to be bored.

"Boredom is not just boring," she writes in *Dopamine Nation*. "It can also be terrifying. It forces us to come face-to-face with bigger questions of meaning and purpose. But boredom is also an opportunity for discovery and invention. It creates the space necessary for a new thought to form, without which we're endlessly reacting to stimuli around us, rather than allowing ourselves to be within our lived experience" (Lembke 2023, 41).

Stealing the Bundt Pan

Ask any early childhood expert how they feel about boredom and their eyes will light up. Boredom, it seems, is the secret precursor to every creative adventure that comes next. It's how elaborate marble runs are built, how cardboard boxes become ice-cream shops. It's why your toddler keeps stealing the bundt pan to use as a busby hat for her teddy bear parade. It's the reason we, even now as adults, have happy memories of playing kick the can with neighbors until the streetlights came on, of losing ourselves in the sheer joy of running through the backyard sprinkler—again and again and again and again.

We weren't just wasting time, we were creating. We were learning. We were experimenting, lost in our own world of imagination and dreams and goals. Or as psychologist Mihaly Csikszentmihalyi calls it in his bestselling book, we were in the delicate state of flow, where "enjoyment appears at the boundary between boredom and anxiety" (Csikszentmihalyi 2008, 52).

But in the digital world, there is no delicate balance. Fast-paced imagery and clamoring graphics plow through the boundary of boredom, crashing into overstimulation and, as a result, anxiety. "The bored self is conditioned to respond to the brightest, loudest, and most shocking volley for its attention," Dr. Kevin Gary explains to me. "Yet to sustain this constant stimulation, each arresting moment needs to be surpassed by yet another and another. The problem, then, is we do not ever cross the threshold which is the other side of boredom."

Dr. Gary introduced me to a something I hadn't heard of before: boredom prompted by overstimulation, what Robert Louis Stevenson called the "weariness of satiety." "This sounds counterintuitive, given that boredom is usually linked to understimulation," he tells me. "But an oversaturation of stimuli hinders the ability to attend to any one thing, [leading to] irritability, restlessness, moodiness, and a desire to be simply anesthetized."

Thus, the cycle continues. Too little stimuli leaves us bored. But too much *also* leaves us bored. So where's the sweet spot? How do we pull ourselves out of passivity, out of mindless and infinite scrolling and into a true state of full engagement, creativity, and presence? How do we invite boredom? How do we turn autoplay into play?

And how do we bring our kids with us?

We follow Big Tech's formula for engagement: spark starting cues, and remove stopping cues.

Spark Starting Cues

TikTok's starting cues capitalize on the idea of FOMO—fear of missing out. There's a party already going on, the app lures, and you have an open-ended invitation. The idea reminds me of a concept I first learned studying the Reggio Emilia approach to early childhood education. The Reggio philosophy surmises that children learn best through firsthand exploration that engages all of the senses. It's a relaxed approach where a parent's role is simply to curate open-ended materials in a creative, nondirective environment. Set the scene and let the child discover it on his or her own terms. The name of the concept? Invitation to play.

Invitation to Play

Putting this philosophy into practice might mean clearing your coffee table during a toddler's naptime, then placing a few loose objects—junk mail, a paintbrush, some watercolors—on display for when she wakes. There's no commentary involved, just patience as she finds and explores the pieces independently. There's no outcome, either. What will she create? It's up to her.

While the concept is highly effective in early childhood, you can easily—and likely already do—put this approach into practice with *everyone* in your home. Invitation to play is the same effect that makes your teen magically appear in the kitchen just as you're pulling pizza out of the oven. It's the reason raucous laughter can make everyone in a room full of people look up from their phones. It's why, if your spouse were to find you blowing up a few balloons on a Tuesday afternoon, he or she would ask, "What's the occasion?"

OPT OUT

"Last weekend, I put out a bunch of mismatched socks that had long ago lost their pairs, a few Sharpies, and some old Christmas ribbon. My daughters, eight and ten, made puppets, which led to the idea of hosting a neighborhood puppet show in a few weeks. They're building the 'set' out of cardboard as I type this."
—*Tammy F.*

An invitation to play—just like TikTok—signals that there's already a party going on and all that's missing is you. And just like that, you've got your starting cue.

So what makes a good starting cue? Anything goes. "The great outdoors is the place most loaded with open-ended materials," Lawrence J. Cohen, the author of *Unplug and Play*, tells me. "Sticks, rocks, walls, creeks, snow, puddles, dirt, trees." Natural elements provide dozens of opportunities for creation—mud into clay, pine cones into stamps, feathers into paintbrushes, stones into canvases. Try displaying a few backyard finds on your kitchen counter and see what happens next.

Or look around your home for items you already have. Large sheets of butcher-block paper and a few markers sprawled out down the hallway floor. A used ukulele. A working puzzle on the living-room coffee table. Root-beer float ingredients on the kitchen counter. An old digital camera. Your kids' favorite band on the Bluetooth speaker. Pipe cleaners. A broken toaster and a screwdriver. Sticks and strings. Recycled boxes and glue. Binoculars and a backyard field guide. A slapdash block tower begging to be knocked over. An old sweater and a pair of scissors. A deck of cards, shuffled and dealt for a rousing game of euchre. An old yearbook of yours, open to a particularly amusing photo.

When planning your starting cue, remember this: display, don't say. Kids are visual and immersive by nature. What sounds good to a child is rarely the same as what *looks* good to a child. This is why our rote suggestions of activities—*Why don't you play outside? Why don't you try your new drawing book? Why don't you play cards?*—never truly motivate. Boredom is a sensory feeling that calls for a sensory response. Consider: if you were hungry, which option would you prefer? A friend saying, "Why don't you make pot roast?" Or a friend placing a fresh-from-the-dutch-oven pot roast right in front of you to eat? Boredom is the same way: we can suggest a hundred different meals, but until we actually eat, we're still hungry.

⬇️ **DM**

"I started incorporating starting cues as a screen-free way for her to transition from the school day. It has been a nice, fun experience for her to decompress and to reintegrate into home life. Her energy has totally shifted."
—Lora P.

Starting cues work because they're unannounced, unforced, and unexpected. Try just one idea a day. Let the objects speak, and let your child explore them on their own terms, in an outcome-free environment. There's no pressure or expectation to make or do anything at all. It's an invitation, not an assignment.

Within a few short weeks, you'll have created a steady rhythm of starting cues to spark what Silicon Valley startups often refer to as the flywheel effect. This phrase coined by Stanford researcher and author Jim Collins (Collins 2001) refers to a retention and satisfaction strategy that boasts three key steps: attract, engage, and delight. It works just as you'd assume it would: An open yearbook on the kitchen counter attracts your teen. He engages with a photo, spotting you in your terrible haircut, ribbed bodysuit, and winged eyeliner. He laughs wildly, delighted enough to share with a sibling, a friend, your spouse—anyone within earshot—and spends the next hour flipping past homecoming floats and school spirit days, wondering aloud why the computers were all larger than a standard microwave.

Boom—momentum is born and the flywheel effect has begun.

LOG OFF

Share with others the magic of the flywheel effect. One grandmother I spoke with found it difficult to connect with her faraway grandkids on a regular basis. So she ordered a book of knock-knock jokes. Now she calls her grandchildren every day at 4:00 p.m. to read a different one aloud. "Sometimes it's just the joke and then we hang up. But most of the time, they have a story to share of something that happened at school, and we gab on and on from there. It's really transformed the relationship I have with my grandkids, and with my daughter, too."

Remove Stopping Cues

Once you're into the rhythm of introducing starting cues and igniting an atmosphere of play in your home, it's time to follow the next step in Big Tech's formula: remove stopping cues.

A stopping cue is anything that pulls or nudges your child out of engaged play, contemplation, flow, or creativity. During hours of conversations I had with leading childhood researchers, experts helped me to narrow down the most popular stopping cues that we as parents, however well-intentioned, often employ with our children when it comes to play and creative pursuits:

Time

"The biggest—and rarest—factor in encouraging children into the flow state of play is enough time," Dr. Cohen tells me. "The deepest experience of play requires hours, not minutes. This will require concerted effort by committed parents and visionary educators to become the norm for children."

Concerted effort indeed. Wide swaths of time feels like an impossible luxury to nearly everyone I know. But investing in clearing your child's schedule as much as possible reaps infinite benefits for the entire household. Here's how:

If Your Child Is Four and Under

You're in luck! Time is exactly what you have with your young children. Chances are, you're not spending your evenings shuffling them to and from Spanish club and swim practice. With your child, the challenge won't be finding the time to play but *using* that time to play. Attention spans in tots are still very short, and as with any skill, independent play takes practice. Because your toddler likely won't be building Hot Wheels tracks by himself for four hours, you have to show him what it looks like to engage in play on a parallel level. Try inviting him into your activities when you can: offer empty pots and pans while you cook, a sponge while washing dishes, a notepad while you're making lists, a stuffed animal while you're rocking the baby. The more your child is encouraged to join you in your duties, the more he or she will get a front-row seat to see what attentiveness and focus look like. To you, it's work. To your child, it's play. After all, in the words of famed educator Dr. Maria Montessori, "Play is the work of the child."

If Your Child Is Five to Twelve

Your calendar looks a bit like a game of *Tetris*, doesn't it? Elementary and middle school years are sticky spots for most parents, because extracurricular activities and school commitments ramp up considerably during this time. Also, the older your child gets, the more drawn to social activities he or she can be. Clubs, sports teams, and community events offer a chance to see and connect with friendships outside of the home.

But as wonderful as structured activities can be, they're just that: structured. They're outcome based, meaning your child isn't getting a chance to explore the same type of creative solutions that independent, unstructured play can offer. Think of the difference between a paint-by-numbers kit and a blank canvas. With paint by numbers, your child is following directions that lead to a fixed result. With the canvas? Your child is in charge, from start to finish. For today's risk-averse children, that autonomy is key. The opportunity for low-stakes failures, to pivot ideas, to self-reflect, to critique their work, to encounter problems and explore fixes? *That's* the benefit of open-ended play.

While most of us can't shred the calendar and start from scratch, there are many ways to prioritize play in a busy family. Can you commit to two Sundays a month when nothing is scheduled, nothing is planned, and white space awaits? Can you choose an evening each week devoted to play—a familywide Lego contest or a bake-off or whatever the kids choose? Is Saturday morning a good time for tree climbing in the woods? Can you say no to unnecessary meetings? Can you decline a few social invitations? Can you keep extracurriculars to the 1:1:1 rule (one thing per kid per season)?

For children who thrive in structure, we can still make room for open-ended play by seeking out short-term commitments.

Rather than signing up for a club or team that meets every evening of the week, keep an eye out for one-off events in your hometown—a kids' cooking class, a beginners basketball scrimmage at the Y, or even free juggling at a summer festival. In a few simple interactions, your child will have been introduced to something new, something engaging, and something to spark their curiosity as they move through this world.

Remember: play is worth prioritizing, both for your child and for your connection to that child. In the words of Zig Ziglar, "To a child, love is spelled T-I-M-E."

If Your Child Is Thirteen to Eighteen

Play. What's that? The older our children get, the less we assume they need it (and us). As they integrate more into friend groups and individual passions, it's natural to feel like their whole world is spinning in another orbit entirely. But it doesn't have to.

For teens, play looks different. It might be trivia night at a pizzeria, a friendly debate over a current event, a quick jog through the neighborhood. Late-night cookies go a long way toward connection and conversation, and so do impromptu road trips. Creating space for them to pursue their passions—whether songwriting or jewelry making or charcoal art or tae kwon do—and investing in their talents are key ways we can still offer an ongoing invitation to play.

One mother I spoke with let her twins build a skate ramp in her unfinished basement. Another made room in her home office so her daughter could have a writing desk to work on a young adult novel. A dad I spoke with spent a summer helping his son restore a Mustang, while another became an investor in his daughter's beekeeping business.

Fostering play in the minds and hearts of our older children looks a lot like taking the time to support them. A good place to start is asking your child, "If you had all the time in the world, what would you want to create?" And then off you go.

Space

The second barrier to play that trips up many of us—and our children—is inadequate space. There are two ways space can manifest in stopping cues: a physical-space barrier, and a mental-space barrier.

Physical-Space Barrier

Physical spaces can create stopping cues when there isn't sufficient room for a project, activity, or game. For many of us, this problem can be solved by venturing to an open field in a park, kid-proofing a living room, or building bunk beds to free up a spare bedroom for a safe and secure play area. But more often than not, there's an even easier fix: a toy purge.

Toy Purge

Many of us assume that more toys lead to more play, right? But research suggests the opposite. "Fewer toys, it turns out, result in healthier play, and, ultimately, deeper cognitive development," says social psychologist Susan Newman (Newman 2017).

Need help determining what toys to keep and what toys to ditch? Here are three tried-and-true strategies:

1. *The stealth purge (best for getting rid of unused items).* Ask your child, "Can I tidy this space for you?" Remove any items you haven't noticed your child playing with and place them in a bin, box, or bag to keep out of sight for a few months. If an item is missed, you can easily recover it from storage. If an item is forgotten after six months, feel free to donate or save for a future child. (Note: Your child's favorite toys are often not your favorite toys! It's important to be as impartial as possible here to ensure that what your child truly loves is front and center. Yes, this includes the Hulk action figure, the megasized giraffe, or the plastic candy-colored Barbie Dreamhouse.)

2. *The primary purge (best for getting rid of excess items).* Ask your child, "Which is your favorite?" For example, your craft drawer might contain dozens of different writing utensils (glitter pens! crayons! colored pencils!), but your child really only likes using markers. Or your basket of blocks might contain multiple sets, but the wooden ones appeal more to your sensory-sensitive child. The primary purge offers insight into which items your child primarily uses and loves, so you can donate the rest with little resistance.

3. *The someday purge (best for getting rid of broken and discarded items).* Ask your child, "Can I replace this for you?" For example, your son might have six foam swords with broken handles. The someday purge opens a dialog about the item: does he have plans to repurpose them, or are they just taking up space because broken swords are better than no swords at all? By inviting your son into the process, you can determine what might spark more play. Full replacements? Fixing the swords? Swapping the foam versions for fewer swords, higher quality swords? Make a plan together.

Play Zones > Playrooms

Now that you've sorted through the toys and items your child enjoys, where will you keep them? For many of us, a perfectly designed playroom is a pipe dream. Space is limited for those with large families or small homes, and many times rental restrictions can limit creative solutions. Don't fret: the truth is, separate playrooms often *discourage* independent play. Children, after all, want nothing more than to navigate life alongside their parents. They want an invitation, "Come play!" rather than a perceived exclusion, "Go play."

With this in mind, consider play zones that are integrated into your everyday activities. A play kitchen near your kitchen will get far more independent use as your child models your actions—prepping breakfast, clearing the table, washing dishes. A simple book basket by the living room sofa is far more likely to be combed through than floor-to-ceiling bookshelves in a play-room (no matter how beautiful!). And loose-leaf paper and crayons in your desk drawer might spark your child to send a letter to Grandmother while you send an email to a client.

By scattering play zones and activities throughout your home, you're reinforcing the idea that play exists within a child, not a room. Creativity can be accomplished anywhere and with anything.

⊙ **DM**

Use your whole house. I know a mother who never uses her micro-wave, so she surprises her young kids by hiding a "secret treasure" in it every morning: a Matchbox car unearthed from the couch, a stuffed animal that has seen better days, a fresh tub of Play-Doh. The kids love popping open the door to see what's inside and almost always scamper off to play with it so she can prep breakfast independently.

Function

To avoid stopping cues, it's important to note the function of a play space. The question isn't, "How does this space feel to me?" but, "How does it feel to my child?" Try this experiment: Drop down to your child's height. Get on your knees if you have to. How does the room feel from this vantage point? Is it inviting? Cumbersome? Intimidating? Cozy?

Function means different things to different humans—kids included! Most children prefer books to be displayed by their covers, not their spines. Does that make the space feel tidier? Not neces-sarily. But is it more functional for the child? In most cases, yes. Likewise, closed storage feels good to most adults but often means a child's favorite toys are inaccessible or left undiscovered.

Consider your child's personality to prioritize what's most func-tional for them, always aiming to establish good habits up front. For example, a child who requires a low sensory input might prefer a credenza or sideboard with a few of the day's favorite toys displayed and the rest hidden away to rotate in at his or her whim. A child who requires a high sensory input might be more accepting of and inspired by open shelving. Partner with the whole family to create a plan that feels empowering and sustainable to maintain.

Storage

When we're making room for a child's creativity, we must also make room for a child's creations. Perhaps you keep a low surface empty for ongoing Lego builds, agreeing that after a set amount of time the creation will be put away and repurposed into something new later. Maybe you clear off an entryway shelf for nature finds, agreeing that after they've been displayed for a week, they can be returned to the great outdoors. And of course, there's always the refrigerator: the art gallery curated daily by parents worldwide.

Whatever you decide, keep this phrase in your back pocket to share with your child: "There's something new to create tomorrow." By providing the encouragement (and the space) to fulfill this mindset, your child will be well steeped in creativity day after day.

⏻ LOG OFF

Feeling buried in kids' artwork and sentimental creations? Finding a way to capture them might make parting with temporary creations easier. I like to keep a secret email address for each of my children. Periodically, I send photos of their sculptures, artwork, and creations to that email address. When they're older, they'll have a treasure trove of memories to enjoy for posterity.

Toys

Last, remember: a play space needn't house a single toy. Items that we live with and use in our daily lives are magnets to children, so much so that child psychologist Elinor Goldschmied coined the phrase *heuristic play*—or object play—in the early 1980s. Because a child's deepest inherent goal is to make sense of the world around them, their brains are like an archaeologist's on a dig: *What purpose does this serve? What does this thing do? What does it mean? How does it work?*

Encourage your child's exploration of everyday items. Let curiosity take over. Open the kitchen drawer and observe as your toddler figures out the salad spinner. Give your second grader the handheld vacuum so he can see how it works. Hand over your cosmetics bag and prepare to be amazed as your six-year-old busies herself for a good half hour. By presenting the unexpected, you're offering your child an open-ended, multidisciplinary, free(!) source of discovery in the playroom and beyond.

Mental-Space Barrier

Once you've removed stopping cues from a child's outside world, it's time to address their inside world. A mental-space barrier to play is any incoming information that interrupts, distracts, or removes them from their play. And like most things, it often arrives with the best of intentions.

Imagine your second grader is outside in the back yard when you notice he's not wearing his coat. Unless it's an extremely unsafe scenario, resist the temptation to interrupt his play by sending him inside for his coat. He's likely fine, and the interruption will most certainly derail the flow of his creativity.

Or maybe your fourth grader is frustrated with her knitting needles and you spot what's wrong. Unless you're asked for assistance, pause and watch what happens next. If she's in the state of flow, she's primed to solve the problem herself. Allowing this sort of troubleshooting only enhances her experience, builds her self-confidence, and boosts autonomy, thus giving her more enjoyment in her play.

In a study called "The IKEA Effect," Harvard researchers found that people value a creation or an experience more when they've completed it—start to finish—themselves. "We show that labor leads to love only when [it] results in successful completion of tasks," the researchers concluded. "When participants . . . failed to complete them, the IKEA effect dissipated" (Norton, Mochon, and Ariely, 2012).

When we don't allow our children to complete the creative cycle themselves and we swoop in uninvited, we not only interrupt the process of play but remove the value of the experience. As difficult as it is, try not to infiltrate your child's play with your idea of fun, your version of the outcome, or your fix for the problem. Stay near, and stay available should your child need your assistance. But until then? Let them be.

Momentum Matters

A gentle reminder: momentum and flow aren't just for our children. When it comes to moving through the world alongside each other, stopping cues are a surefire way to short-circuit any connection. Stay mindful of the many ways we do this with the people closest to us. How often have we stopped a conversation to google the name of that one band we can't think of? Or interrupted our brain's attempt to describe a place or event by pulling up a photo instead?

Catherine Steiner-Adair, clinical psychologist and author of *The Big Disconnect*, writes, "Every time our child's texting, TV, electronic games, and social networking take the place of family, and every time our tech habits interrupt our time with them, that [connection] is broken and the primacy of family takes another hit" (Steiner-Adair and Barker 2014, 41).

Momentum matters—both online and offline. Resist the temptation to invite a roomful of tech programmers into the limited time you have with your children. Every time you decide not to search a fact, not to google a solution, not to pull up a photo, you're teaching your child that screens interrupt life, not the other way around.

OPT OUT

In our home, we challenge our children to tell good stories without visual aids. So when someone wants to show a picture of their prom dress, puppy, or vacation, we like to ask if they'll describe it to us instead. Nine times out of ten, their description paired with our imagination is far better than the photo itself.

This evening, I called the kids into the kitchen for dinner. Both girls arrived quickly, sweaty and famished from a rousing game of freeze bounce on the trampoline. "Where's your brother?" I ask.

But then I hear the opening chords to "Poor Unfortunate Souls"—C major—and I know right where he is.

"Just a minute!" he shouts over a string of Es.

And I say yes.

 SELFIE

Think about the last time technology interrupted the momentum of your day. What boundaries can you put in place to ensure creativity and flow and connection are protected in your home? What starting cues might you add? What stopping cues might you remove?

Remember: tech-free doesn't always mean no-tech. It can mean, simply, free from tech. It can mean allowing yourself the time and space to consider solutions that will release your family from tech's influence—at any moment, for any reason. How might you build a household that's liberated from the addictive nature of technology? Get specific as you explore and envision the infinite possibilities available to you and your unique family.

Chapter 9

Alexa, Meet Baby

Reevaluating the Digital Path We're Paving

All men make mistakes, but a good man
yields when he knows his course is wrong,
and repairs the evil. The only crime is pride.
—Sophocles, 496 BC

Train up a child in the way he should go;
even when he is old he will not depart from it.
—Proverbs 22:6

TECH'S PLAYBOOK

> 🔘 The conveniences of technology will make your life easier.

✅ **OUR PLAYBOOK**

> 🔘 Easier is not better.

Today, here in the Midwest, a burst of sunshine greets us in the middle of a long stretch of winter's gloom. For a single afternoon, the neighborhood comes alive—all manner of bikes and skateboards and scooters whoosh by with the familiar scratching sound their wheels make on the pavement. After dropping my older two children at their theater troupe's rehearsal, the toddler and I take to a nearby park.

We swing and chase and run and climb. She scuffs her knees, finds a crocus, pats a baby. We share bananas and watch the clouds. By all accounts, a perfect afternoon.

As we prepare to leave, I watch her plop into the passenger's seat of the playground car next to a preschool-aged boy manning the steering wheel. "Directions to Blake's house," he says as he drives, stating his requests into a pretend phone he's brought for the occasion.

I smile, but he doesn't acknowledge me.

He continues. "Siri, call Blake," he commands. "Send Cruz a text. We are running late, period. We will still come, period. We are coming now, period."

The boy calls Blake again, then sends another text to Cruz. "Do you want pizza, question mark. I will bring pizza, period." He puts his phone down to drive, picks it back up. "Siri, order a pizza," he says, then starts to type furiously on his "phone," lost in a land of his design. I am not surprised; I look up and see every other adult here doing the same.

When it's time to go, I say to my toddler, "Excuse me, ma'am. I think this car is out of gas." She feigns concern but knows it's all a ruse to get her to pack up and head home. As we walk to the parking lot, I hear one final command from the boy at the playground.

"Alexa, get gas."

The Chirp of Technology

Stroll through any public space and it's clear how much smartphone technology is carved into the landscape of our society. QR codes instead of menus. App rewards instead of paper coupons. Bar codes instead of credit cards.

Even in our local botanical gardens—a place where families can disconnect to surround themselves with cascading waterfalls, orchids and palms, cacti and tropics—Big Tech is blooming. A recent renovation to the children's center removed a play kitchen, tables, and books. In its place? A life-size interaction screen.

"The last time we visited," says a mother I spoke with, "[my son] didn't even want to see the gardens! He was too mesmerized by the screen."

Even when we arrive home from our tech-laden outings, many of our private spaces greet us with the chirp of technology. Google Nest thermostats keep us warm—and know when we're gone (Google, n.d.). Philips Hue lighting keeps us out of the dark—and open to hackers (Hollister 2020). Alexa—our nation's favorite "virtual assistant" used by 70 percent of smart-speaker owners in the US (SafeAtLast 2022)—plays music, makes to-do lists, sets alarms, streams podcasts, plays audio books, provides weather, traffic, sports, and news updates. It does nearly *everything*. And listens as we do nearly everything, too.

"I was having a really honest conversation with my husband," one mother tells me. "I was feeling kind of invisible, like what I was doing here at home was underappreciated or unnoticed. It was just a really raw, truthful moment. Really beautiful. And then I was interrupted by Alexa's voice recommending a suicide hotline for me to call."

Amazon would argue Alexa was doing what her AI technology was designed to do: spit out predictions and information gathered and filtered through its algorithm. "Alexa makes your life easier, more meaningful, and more fun by letting you voice-control your world," Amazon touts on its sales page for the AI technology. "Alexa can help you get more out of the things you already love and discover new possibilities you've never imagined" (Amazon, n.d.).

But not all new possibilities are worth imagining. One mother stood with mouth agape when Alexa responded to her bored ten-year-old asking for a challenge. "Here's something I found on the web. . . . The challenge is simple: plug in a phone charger about halfway into a wall outlet, then touch a penny to the exposed prongs" (Clark 2021).

Parents all over the country report unsettling information dispensed by Alexa—countless tales of inappropriate bedtime stories, glaring disinformation, misunderstood requests. On a particularly disturbing YouTube video I found, a small redhead boy lispily asks Alexa to play the song "Tigger Tigger." Alexa, confused at first, soon offers "pussy anal dildo ring" while his parents shout for Alexa to stop. The family clip ends with the parents bursting into laughter at the sheer humor of it all. But the boy smiles uncomfortably, unsure whether he's in on the joke (F0t0b0y 2017).

Artificial intelligence might offer intelligence, but it's just that: artificial. Contrived, counterfeit. False. Unnatural. *Unreal*. For a society that praises authenticity and truth, it's puzzling how quickly we've raced to welcome a technology that offers the opposite of what many of us claim to value.

So how did we get here? Why have we invited Big Tech into our homes and schools and lives and dinner tables and cars and sports and *world* without weighing the cost of what arrives along with it? Why are we allowing smart devices to interrupt our days with notifications, pings, updates, reminders? And why are we *paying* them to?

If you're a Silicon Valley venture capitalist, you already know the answer.

First-Order Thinking

"We've had a high-tech consulting agency for twelve years now," Silicon Valley technologist Dr. Drea Burbank tells me in an interview. "Over time, we started to see that people would come to us for consulting, but what they really wanted was to learn how to live in a world that was technologically changing too fast. We began to overtly train thinking, instead of skills. The thinking [comes] first. . . . Silicon Valley has developed life skills that should be widely applied outside the context of technology, [creating] profound implications if the principles are clearly understood."

One of the more profound skills I've learned from Silicon Valley is first-order thinking versus second-order thinking. First-order thinking might look like this: *My child wants a smartphone. He's thirteen. The surgeon general recommends waiting until thirteen* (Gordon and Brown 2023). *I think he's ready. Let's buy him a phone.* But second-order thinking asks, *And then what?*

How much time can you dedicate to setting up parental controls and apps? How will you address cyberbullying, porn? Will you allow social media, and when? Which apps are okay to use? Which apps aren't? What are your screen-time rules and guidelines? What sites can he access, and when? What parameters are in place at school, in a friend's home? Which contacts can he add to his phone? What consequences will be in effect for crossed boundaries? How much time can you offer to monitor his texts? How much screen time is okay? What are the best rules for your child?

And what are the best rules for *you?*

"Everyone talks about how much time kids are spending on their phones, but no one told me how much time *I'd* be spending on his phone just to stay on top of it all," one father told me. "I'd think Google Docs was fine, but then I'd hear from another parent at school that there's a shared

DM

"I almost bought a smartphone for my ten-year-old and twelve-year-old because their team sports and school info and club meetings all ran on separate apps and it was so much to keep track of. But then I realized that I can either put in the time sorting out schedules, or put in the time sorting out screen distractions, limits, boundaries, and placing all kinds of barriers between us. I chose schedules. Now we're an opt-out family."
—*Cami W.*

spreadsheet with nude pics going around. Or they'd just switch hot spots so they could get on whatever sites they wanted. Changing time zones so they could spend longer on their devices, doing factory resets to clear an app's data. I mean, kids know exactly how to get around this stuff."

I asked him what the positives to a smartphone have been. "You know, I'm sure they're there," he says, "but it's hard to see that right now. Devices have just made our lives so much more complicated."

But if Silicon Valley has shown us the way in, it also shows us the way out.

Traction Map

"Traction maps are about looking at hard evidence of what matters as a way to measure progress," Dr. Burbank tells me. "We do this in the Valley because sometimes when you are in the thick of [it], it's impossible to tell how you are doing. Your senses are blinded, you are disoriented, distracted, everyone is talking all the time. How do you orient and pull through it? You make a traction map of what really matters, and you just focus on that, every day, one step at a time no matter what chaos surrounds you."

A traction map says where you want to go and how you want to get there. Along the way, it identifies specific points, actions, and pivots that will either keep you on course or distract you from your greatest purpose. So what does our traction map look like—as parents, as CEOs of our homes? What proof do we have that devices are moving us forward into a better place, both individually and as families? Does texting help us better communicate? Does social media strengthen our bonds? Do we like where our traction maps end?

Or how about where they *start?*

Many parents I spoke with mentioned how quickly they integrated various technologies into their homes without even thinking about it, especially at the beginning. "There are all these apps that promise to help make parenting easier," one mother said. "It starts with pregnancy—you've got, like, estimated due dates or baby registries or even contraction timers. And then your baby is born and, by that time, you're already all-in."

I asked her to show me how many baby or parenting apps she has for her seventeen-month-old, and she unlocks her iPhone screen. Four swipes later, I'm staring at a conglomeration of diaper-delivery services, sleep apps, sound machines, cloud baby monitors. "I don't use them all," she says. "But definitely this one."

She's talking about Baby Tracker. "Baby Tracker offers a simple, streamlined way to track your baby's daily habits, health, and exciting 'firsts' of those precious early days and months," the app's description proclaims. "Record feedings, diaper changes, and sleep patterns with a quick one-handed tap, then feel free to go back later and add details and even photos."

She tells me that Baby Tracker uses Siri to input her daughter's medical information, her first smile, her favorite brand of formula. According to the terms of use (Big Brains 2022), "2. Baby Tracker may obtain certain personal information you provide when you use the sync feature with Baby Tracker Server. Such information includes but is not limited to the following: (a) your baby's name; (b) your email address; (c) your baby's date of birth; (d) your baby's growth data; (e) all records that you input in Baby Tracker; (f) all photos that you input in Baby Tracker. . . . 5. We will retain your data for as long as you use Baby Tracker and for a reasonable time thereafter."

"It's, like, the ultimate baby book," she says. "It's so nice, because parenting is really hard, and anything that simplifies that is going to free up my time to engage with [my baby]." Her baby stirs, and she lifts a finger to pause our conversation. "Sorry," she says. "Just need to log her naptime real fast."

Experiencing parenthood—especially for the first time—can feel scary, and many parents alleviate their fears with information, data, and statistics. But many pediatricians and psychologists argue that more data doesn't lead to more confidence. It's often the opposite. "Monitoring isn't helping alleviate our anxieties," founder of Family in Focus Dr. Wendy Schofer tells me. "It actually increases it. The monitors activate our fears that there is something to keep an eye on. [And] the mere tracking can be problematic because of the mental burden that it places on the parent doing the tracking. It's creating more work and bandwidth to track. The focus that [parents] place on monitoring the child's stooling, behaviors, and milestones can detract from the time that they get to spend being present with their children. Or *sleeping*."

For new mothers, Big Tech promises certainty, facts, measurables. But at what cost? "Our kids and their behaviors are not meant to be quantified," Dr. Schofer adds. "While I can't speak to the existing third-party-advertiser use of data from monitoring apps, that should be enough of a reason not to invite extra entities into the home."

One mother I spoke with talked about how she felt like tech brands prey on her fear as a new mother. "I just kept hearing these messages that if I wasn't actively engaging [my baby] with music or content or whatever, then he wasn't going to be developmentally on track," she says. "We did it all: the tummy time apps and the diaper change trackers and, like, BabyTV. But now I kind of wish he wouldn't have grown up with all that. He's glued to my phone at restaurants now. He's always asking for more screen time. I think it just kind of taught him that these apps are always going to be there, that they're just part of his daily life."

Those baby apps, she says, were the gateway drug.

"My first child was born in 1995, long before the advent of the smartphone," Jen Pepito, author of *Mothering by the Book*, tells me. "Our early days as a family were characterized by lots of time together, exploring the natural beauty surrounding us. . . . I didn't know about attachment theory or how important eye contact was, [but] I did my very best to be present. I loved seeing the light in my children's eyes as I approached them for a feeding, and it was a joy to be their mom. I'm so grateful that smartphones weren't available in that season, because even with my intention to be present with my children, once I had a phone, it's been a struggle to keep it in its proper place."

Big Tech knows the strategy. If they can get past the gatekeeper, they get everything thereafter.

Clinical psychologist and author Catherine Steiner-Adair addresses this path in her book *The Big Disconnect*. "When that training starts early, as it does now for young [kids], they get so used to it that, unlike the newborn baby who innately knows something is missing and complains about it, our older tech-trained children don't even know what they have lost," Catherine says. "Just because your baby can tap a touch screen to change a picture does not mean that he should, that it is a developmentally useful or appropriate activity for him" (Steiner-Adair and Barker 2014, 79).

In fact, it's not. "Engaging with [screens] may itself be rerouting brain development in ways that eliminate development of essential neural connections your child needs to develop reading, writing, and higher-level thinking later," she adds.

⊽ DM

"There are so many modern kids' shows like *Octonauts*, *Wild Kratts*, and *PJ Masks* that feature main characters with devices. I didn't like the message that was sending, like my kids couldn't do cool things without having a device strapped to them. Once we stopped watching them, my kids stopped asking for smartwatches. I don't think that's a coincidence."
—Lisa E.

But by that time, the traction map is already well underway. BabyTV grows into ABCmouse, which grows into YouTube Kids and FaceTime and Facetune and far beyond.

Then what?

We look, once again, to Silicon Valley's venture capitalists. Dr. Burbank walked me through many of the Valley's standby tools, but one sounded especially familiar. Famed author and educator Stephen Covey calls it beginning with the end in mind. Zen Buddhists call it *memento mori*, "Remember you must die."

Silicon Valley calls it an exit strategy.

"When Silicon Valley asks a founder to [consider] their exit strategy," Dr. Burbank explains, "what they are really saying is that life is never forever, and have you really thought about where you are going?" Or more aptly, when and where and how to get out? "They want to see if the founder has the maturity to see that the future is a branching tree, and that while there is inherent uncertainty, there are also endings and beginnings, and you should think of the ending at the same time as you think of the beginning."

Our job is clear: it's time to consider an exit strategy.

If we're questioning whether our teens are ready for a smartphone, why are we also not questioning when they—or we—might be ready to abandon it?

Seventy-three percent of Americans report they'd be happier without their phones (Moller-Nielson 2022). "To be honest, my relationship with [technology] is one of the most stressful areas of my life," writes one mother I spoke with. "I feel like I'm being held hostage by it. . . . I just want to live my life without a phone in my hand all day and without people having the expectation that they can reach me 24-7. Unlike many of my younger friends, I've actually experienced that life, and I enjoyed it. I miss it."

Our smartphones nag. They interrupt. They notify, remind, advise, caution, prompt, prod, suggest. They are the equivalent of the overbearing babysitter every one of us as children begged our parents to abandon. *When can we stay home without the sitter? When can we be left in charge? When can we spend the evening alone?*

⏻ LOG OFF

Want to hack your own tech use? Put your iPhone in black-and-white mode, or grayscale, by visiting Settings / Accessibility / Display and Text Size / Color Filters and toggling to select Grayscale. The effect makes your phone less appealing when surrounded by the vibrant, real, and colorful world all around you.

We have stopped asking those questions, both of ourselves and of our children. Instead, we have falsely equated smartphone use with responsibility. "When you're responsible enough," we tell our children, "you can have your own device." Once it's "earned," we hand over a smartphone under the guise of privilege, freedom, trust. But trust takes time to build, as does responsibility. Neither are available at a Verizon store for $50/month.

Tell me, which offers more freedom: allowing your teen to go to the concert, to the museum, to the festival, on the road trip? Or allowing them to go, but only with a built-in, 24-7-monitoring GPS tracking device?

It seems to me that the real privilege, the real empowerment, the real freedom? It's allowing our children to live beyond the babysitter.

OPT OUT

In need of time-tested screen-free swaps for smartphones, smart speakers, tablets, and every other piece of modern technology in your home? Head to *optoutfamily.com/more* for an exhaustive list of alternative ways to live a full and fulfilling life without smart technology.

It's a 24-7 Job

Many parents I spoke with take a graduated approach to tech use, much like a Graduated Driver Licensing system, which helps new drivers gain skills under lower-risk conditions. Kids are given more privileges as they move through various stages. "We started with a flip phone," says one mother I spoke with. "But they weren't cool. My son kept getting singled out. So I switched to a Gabb phone—no internet—and that looked better, but there were just too many things it couldn't do. My son would get put on a group text and half of it would be blocked out because of links or inappropriate content. He wouldn't be able to follow what anyone was talking about. Pretty soon his friends just stopped adding him to the group."

Jean Rogers, director of the Children's Screen Time Action Network, knows firsthand the influx of inappropriate content kids are exposed to once they enter Big Tech's playground. "Even once our children are thirteen, fourteen, and fifteen, we certainly want them exposed to more challenging information about science, nature, pop culture, music, art, and politics," she says. "But why would we ever want them to be exposed to violence, cruelty, and self-harm information? That's not a destination for any age."

Many experts agree. "Personally, as a parent and as a mental-health professional, I have trouble finding benefits that outweigh the risks," says Jillian Amodio, clinical social worker, parenting expert, and founder of Moms for Mental Health (Singer 2023). "Sure, there is some uplifting and inspiring content, but that doesn't negate the content that is more damaging. There is of course the potential for fitting in and achieving social status, but at what cost?"

One father put it this way: "I'm spending more time with my daughter's phone than with my daughter. There's something really wrong with that."

Time and time again, parents cite communication with coaches, employers, or clubs as the main reason their child needs a smartphone. "They've got group texts going for everything," one parent says. "I can't afford to let my kid miss out on important scheduling changes." But when I pressed a few coaches and employers to see whether they'd be willing to send out updates via email or phone calls, they were all happy to accommodate. Even the ones you'd least expect.

⊘ **DM**

"Every day I'm fielding more emergencies, more scares, more stuff they shouldn't be seeing when they're this young. If you're in, you're in. It's a 24-7 job. I think that's why so many parents just kind of give up. There just aren't enough hours. If I could do it again, I'd opt out. All the way, no question about it."
—Maria M.

"As the owner of a Verizon store, my business does rely on mobile apps in some ways," Wireless Zone franchise owner Travis Bauman tells me. "But an employee could absolutely do this job without owning a personal mobile device. In fact, even if I had to do a little more paperwork or legwork, I would happily hire someone who didn't have a mobile device. It is a constant struggle keeping employees working while they are on the clock. Personal mobile devices exacerbate this time-theft more than any other distraction. I would have to imagine an employee who doesn't have a mobile device would have other priorities that would also align with a good work ethic. That may not be the case, but as readily available as devices are now, it would certainly have to be a conscious choice to not have one."

Another local business owner echoed Travis's sentiment. "Oh, I'd hire that person in a heartbeat. But do they even exist?"

We needn't place the blame on ourselves, and certainly not our kids. This is the world we've offered them. This is the one they've got. Former Google engineer James Williams argues that shaming ourselves into a kind of digital detox isn't "the solution, for the same reason that wearing a gas mask for two days a week outside isn't the answer to pollution. It might, for a short period of time, keep certain effects at bay, but it's not sustainable, and it doesn't address the systemic issues" (Hari 2022).

So what does Silicon Valley do when things just aren't working? How does it address its own systemic issues? Suddenly and collectively, the world found out on March 10, 2023. In the second-largest bank failure in US history, Silicon Valley Bank swiftly collapsed after what Danny Moses—the investor who predicted the 2008

> ## ⟳ OPT OUT
>
> **Talk to your child's coaches, leaders, and extracurricular administrators to see whether there are alternative lines of communication in place for an opt-out family. If not, suggest a few! (Find plenty of ideas on *optoutfamily.com/more*.) Likewise, encourage your teen to brainstorm alternative, low-tech paths of communication and scheduling with future or potential employers. Practice making and receiving phone calls from a young age. By being proactive and prepared with solutions, your kids will stand out as capable and responsible employees, ones who can get a leg up without a digital footprint.**

financial crisis in the book *The Big Short*—calls "complete and utter bad risk management" (Helmore 2023).

Two days later, the federal government stepped in to take over tech startup's banking pillar and ensure accounts would be protected. "Thanks to the quick action of my administration over the past few days, Americans can have confidence that the banking system is safe," President Joe Biden announced in an official address (Biden 2023).

The money is safe. Are we?

For parents like Sharon Winkler, there is no bailout in sight. "My seventeen-year-old son, Alex, was coached and encouraged by anonymous social-media users to end his life. He [did]," Sharon tells me. Now an active member of the Screen Time Action Network's Online Harms Prevention work group and Fairplay, she is hopeful in advocating for federal and state internet safety legislation.

Joann Bogard, the mother of Mason, who passed away trying a social-media blackout challenge, has joined Sharon in advocating for change on all fronts. "I was in DC last year with ten other grieving moms lobbying with Federal Senators and Congressmen/women," she tells me. "We have met with presidential assistants, FTC, CDC, special committees, organizations, media, and so many more. The result was that they all wanted change . . . but couldn't agree on the specific changes needed. So this year, more laws are being written, more meetings and hearings are being planned, more prayers are being sent up."

But as of today, risk management still seems to fall on our shoulders as parents.

Roll Up Our Sleeves

This morning, I heard from a friend that our school district has joined more than forty other districts in a lawsuit against Big Tech: Facebook, Instagram, Meta, TikTok, Snapchat, Google, and YouTube. The 281-page claim states their apps are "addictive, damaging to students' mental health, and causing adverse impacts on schools and other government resources" (Arundel 2023).

But according to experts, "major litigation like this is likely to take many years to resolve" (Graham 2022). Until then, we can roll up our sleeves and get to work.

We can reject the idea that our world must be designed for maximum technology use. We can call our favorite diners and ask them to offer paper menus instead of QR codes. We can write letters to the airlines requesting that they waive the fee for printing boarding passes. We can make donations to our favorite advocacy groups. We can raise funds to have our schools test a device-free program using Yondr pouches or phone lockers. We can write to the botanical gardens downtown and volunteer to design and build a children's potting station to replace the Obie gaming screen. We can request that our library offer an offline version of its summer reading program using paper charts instead of an app to track minutes read. We can call our kids' camp director and ask them to reconsider using facial recognition software for minors. We can pass around books and ideas and dreams. We can challenge. We can rally. We can fight.

We can refuse to run headlong down the digital path that's been carved for us. We can do what Big Tech rarely does: We can pause. We can look up. We can turn around. And we can opt out.

➕ +FOLLOW

For a comprehensive list of nonprofits, organizations, and advocacy groups who are fighting against the digital harms of technology, visit *optoutfamily.com/more.*

It will take years. But as Gaia Bernstein, law professor and author of *Unwired*, posits, there was a time when we could never have imagined walking into smokeless bars, restaurants, bowling alleys. But now that is our reality. "The same could happen for a better-balanced tech future," she writes, "once we decide to shift from internal battles to collective action" (Bernstein 2023).

Sounding the Alarm

Years ago, we are given a Kindle Paperwhite Kids tablet for our houseful of readers. We are preparing for upcoming getaways and assume an e-reader will be the perfect solution for on-the-go books. Ken sets up the suggested parental controls, pass codes, PINs, age profiles.

We download a few favorite books and see recommendations for more. But as Ken and I sift through these options, we see an overwhelming number of teen novels and graphic covers—haunting ghosts, bleeding zombies, dark witchcraft. As we hover over the titles, we notice that nearly all fall well outside the age restrictions we've set up.

My husband chats with customer service once, then twice, three times. *Don't worry! You're in charge! Simply search for any specific item and you'll see the option to remove it from your suggestions!* But how do we know what to search for until we see it? How do we know what Amazon will suggest before it's suggested? Can't we remove suggestions altogether? And why is a three-year-old being recommended *Snow White, Blood Red*? We troubleshoot. We research. We call our friends who are well versed in tricky interfaces and deep patterns. There is no fix, we find. So we join our voice to the chorus of many parents writing, calling, reviewing, filing complaints, sounding the alarm.

We grab our helmets, bike to the library, and leave the Kindle behind.

✏ SELFIE

Changing your household's relationship to technology is not easy. But if you feel like you've made a mistake, if you're carrying regret, if you're unhappy with your path and are wishing you'd done it all differently, why wait? You can start today. Make this your family's second act. It's never too late to try a different way. Jot down a few course-corrections you can make as you consider raising opt-out kids, and then begin again.

Chapter 10

Network Effect

Establishing Your Child's Inner Circle

I live my life in widening circles that
reach out across the world.
—Rainer Maria Rilke

And let us consider how we may spur one another
on toward love and good deeds, not giving up
meeting together, as some are in the habit
of doing, but encouraging one another.
—Hebrews 10:24–25 NIV

❌ TECH'S PLAYBOOK

> Smartphones connect us to the whole world!

✅ OUR PLAYBOOK

> ◐ Smartphones disconnect us from the people we're with.

We are at a burger joint with our kids' theater troupe, celebrating a successful run of *Willy Wonka and the Chocolate Factory*. It is late, and raining, and we have just arrived. Ken drops us off and parks the car, and the kids and I run with giddy anticipation as we approach the doors, where lively chatter and bustling energy spill out from the diner.

Inside, we wipe raindrops from our foreheads and weave through the crowd—thirty or so kids, plus their families—to spot a few open seats. We plop down our bags and make our way up to the counter to order, table hopping along the way as we congratulate Charlie, Violet Beauregarde, Augustus Gloop, and his mother.

At last, we return triumphantly with a tray full of fried food. Ken distributes chicken sandwiches, fries, burgers, ketchup—*I forgot ketchup! I'll grab ketchup! No, me!*—straws, napkins, extra ketchup, and together, we pause to say grace amid the chaos and joy.

As I take my first bite, a phone is plopped facedown in the center of our table, then another. Soon three more. Four. Six. I look up to see a circle of actors from the high school troupe who, before flitting about the room to socialize, have come to "turn in" their phones to Ken. I silently raise my eyebrows at him, and he winks.

"Just something we're doing," he says and dives into his food.

I will learn later that he has challenged the group to have a conversation with each other that doesn't involve their phones. They circle around a table in the center of the restaurant, chatting about fear and melatonin, *Brigadoon* and driving. They are still lost in conversation hours later, when the toddlers have all melted down and the younger kids have fallen asleep in booths and the parents have shuffled quietly off for home and bed.

As the diner hushes, the lights dim, and the employees begin to sweep—closing time—one of the girls tells me how weird it feels to be with her peers, no phones at all. "I'm used to that at family dinner, but it's different with friends," she says. "It was awkward at first. We had to find things to talk about. But once we did, it felt like we could talk for hours. We weren't sort of waiting for the other person to stop talking so we could check Snapchat."

Another friend agreed. "Sometimes—and this is terrible, I know—I'll, like, count the heads of people around the table and if there are more than half on their phones, I'm gonna get on mine, too. It's, like, permission? Or something?"

Blue check marks. Going viral. Verification badges. It is no longer enough to be well loved by few. We must be well liked by many. We must hope for popularity, and we must have evidence of said popularity. Likes. Hearts. Follows. Five-star reviews, in both sweaters and friendships.

Social media is built on the foundation of a deeply rooted psychological bias: implied trust in the masses, or—as Silicon Valley claims—social proof. It's why we buy the most recommended vacuum on Amazon, why our daughters add Kylie's lip kit to their Christmas wish list, why our book clubs will read pretty much whatever Maureen Corrigan tells us to.

It's also, according to argumentation theory, fallacious. Popular does not always mean good, and the assumption that it is can often lead us astray. On a small scale, equating social proof with trustworthiness can lead to parasocial relationships and unhealthy attachments. On a large scale? Rampant misinformation on a global level.

Few people understand this more than Neal Mohan, YouTube's chief product officer. Every three months, YouTube removes nearly ten million videos that, according to the platform's safety policy, "can directly lead to egregious real-world harm." In a recent statement, Mohan remarked, "The most important thing we can do is increase the good and decrease the bad" (Mohan 2021).

But who determines what's good and what's bad? What's true and what's not? What's social proof in a world where popularity is sold to the highest bidder?

The Blue Check Mark

Back before Instagram was Instagram and while its cofounders were still hitting the books at Stanford, I was a full-time blogger living in

Los Angeles. At the time, social media didn't yet exist. We couldn't just hop online and share the yarn-bombing installation we knitted on Sepulveda Boulevard or the memoir we'd just published. So I and thousands of other entrepreneurs created our own platforms. We launched blogs on cheap servers, distributed zines we copied at Kinko's, sold ads in the sidebars of our entirely reader-supported websites. We were doing what we loved, chasing passion for pennies. We were young. We were scrappy. We were tired and also deliriously happy.

And then Google AdSense approached the scene—unsurprisingly, with YouTube's Neal Mohan himself at the helm. The promise was this: *We'll do the work for you! Just put this code on your website and never worry about selling ads again. Leave it to us and wait for the checks to roll in.*

Some of us took the bait. Some of us became rich without asking too many questions. Many of us wondered whether the AdSense code was boosting our search ranking, leading more people to our sites (and Google's ads) than ever before. The rest of us wondered who was paying whom, and why. What was the product? The site? The ads?

The *people*.

Just a few years later, I'd meet Jaron Lanier—the computer scientist who is considered a founder in the field of virtual reality and now rallies against the destructive harms of social media—at tech's premiere industry conference SXSW. Over chickpea tacos in Austin, his words registered as a clear forewarning to me: "We cannot have a society in which, if two people wish to communicate, the only way that can happen is if it's financed by a third person who wishes to manipulate them."

Online, that manipulation is rampant. Reviews are often falsified, quotes are faked, celebrity and influencer endorsements are photoshopped on the regular. Small businesses often turn to bots to autocomment on their social-media posts as a way of making them appear to be more popular than they are. Influencers, who are paid per post based on popularity and engagement, can get higher sponsorship deals by hiring a bot service to manipulate their follower count. And many do. One bot service was uncovered earlier this month, "revealing that almost half a million of its fake accounts are being used by high-profile influencers" (Okunytė 2023).

So how do we know what's real, what's fake? Twitter rolled out a solution in 2009—a blue check-mark icon that vetted and "verified" influencers, celebrities, and industry leaders. If you had the blue check, your account was legitimate. The symbol quickly gained traction, and Instagram followed suit, granting verification status to well-known and highly searched-for profiles. But now you can buy your way into authenticity, legitimacy. With Meta Verified, Facebook and Instagram users can pay Zuckerberg a monthly fee of $14.99 per month to get a blue check mark next to their names (Grothaus 2023).

The result? We've seen it. We've lived it. From politics to pandemics, our nation's ethos has been shaped by a faceless algorithm that feeds us whatever it deems most popular.

And it's far easier than we think.

To prove it, sociologist and Princeton professor Matthew Salganik created a music sharing platform with built-in popularity rankings (Salganik 2006). One central question for the experiment was this: What might happen if he tinkered with the ratings, fudged the numbers? What might become of the least popular songs—the ones with the lowest ratings—if he made those rankings appear to be higher than they actually were?

Surprise, surprise. They became *popular*.

The Network Effect

Silicon Valley calls this phenomenon "the network effect." It's the idea that a product or platform increases in value as more people join it—or the more popular it becomes. And the more popular it becomes, the more people join. It's a spiral of growth. Uber isn't as valuable without access to 5.4 million drivers available to drop you to and from any airport in the country. Airbnb isn't as useful if only four rental units are available in, say, Wichita, Kansas. And how fun is your child's smartphone if he has no one to call or text?

The network effect's playbook is this: The product or service relies on its users to generate more money, more time, more attention. And we—the users—are perfectly positioned to line the pockets of Big Tech.

"My son was the last one in his class to get a smartphone," says one mother I spoke with. "We held out as long as we could, but his teacher assigned a school project where the kids had to create Instagram accounts for famous people in history. I just kind of felt like his lack of social-media awareness was getting in the way of his education, you know?"

This is how the network effect works: once the momentum takes hold and reaches critical mass, it's often nearly impossible to imagine life before its influence. We stop questioning whether something is good and right; it's just *there*. Rarely do we see the bigger picture—or what's at stake—until we take a closer look.

OPT OUT

Consider gathering two or three likeminded families to raise opt-out kids alongside you. Set boundaries that feel amenable to everyone, such as no personal devices, no tablets in the car, no gaming. Then brainstorm what replacement activities you can all offer your kids instead. Weekly ice cream socials? Monthly potlucks at the park? Concert tickets? Museum passes? National park memberships? Take turns engaging your kids in new, vibrant experiences—unplugged, yet plugged in to each other.

"Uber, the world's largest taxi company, owns no vehicles," writes *Wired* cofounder Kevin Kelly in his book *The Inevitable: Understanding the Twelve Technological Forces That Will Shape Our Future*. "Facebook, the world's most popular media owner, creates no content . . . and Airbnb, the world's largest accommodation provider, owns no real estate. Something interesting is happening" (Kelly 2016, 109).

It's the network effect. And it's growing at the hands of our kids.

"He's a Phone Person Now"

"My son's middle school teacher keeps a hanging shoe organizer—you know, one of those plastic sleeves with compartments for flip flops?—on her door for every kid's device," says one mother I know. "It's labeled by number, so each kid just drops it in when they come in for the day. It's actually how the teacher takes attendance."

I asked what happens if a kid doesn't bring their device to school or if they don't have one.

"Oh, they have one," she says.

I called my friend Jess to see whether this was her experience in Brooklyn. "It's so sad," my friend says. "[My son and I] will walk home from school and he'll talk about his day and I'll notice he hasn't mentioned playing with his friend Casper or Sam or so-and-so in a while. And I'll ask him about it. He'll say, 'Oh, he got a phone. He's a phone person now.'"

She tells me about a weekend away for her son's soccer tournament. "It's a big highlight for the team and the parents," she says. "These kids have known each other for years; it's this massive celebration. We go out to dinner, get a hotel with a pool, cause a ruckus—it's just always such a blast."

But this year, it was different. "This was the summer after sixth grade, and the kids all got phones," she says. Her son and his friend—who are both opt-out kids—swam in an empty hotel pool, practiced cannonballs, talked about the upcoming game. After a while, the rest of the teammates came downstairs from their rooms. Tucking their phones under their towels, they'd jump in the pool for a while, then dry off and get out of the pool to check sports stats or get back to their video games. "Just dipping in, dipping out," my friend said. "Like users!"

But sometimes users quit.

"Just as network effects create a rich-becomes-richer cycle leading to rapid growth of the network, reverse network effects can work in the opposite direction, leading to users quitting the network in droves," says Sangeet Paul Choudary, a global CXO and digital advisor (Choudary 2014). The idea is simple: once the platform or product grows beyond its capacity to deliver its promise, there's a breakdown in trust. From there, collapse is imminent.

Choudary explains that the health of most successful networks hinges on maintaining a balance of three main tenets: connection, content, and clout. If a network grows to an unsustainable scale, the level of enjoyment users experience in these areas plummets. Think: More users leads to less connection by creating more spam and solicitation requests to sift through. More users leads to less meaningful content, reducing the ability to personalize and curate a user's experience. More users leads to less clout—or credibility—relying on algorithms to spotlight power users who might not be the best or the first but are the loudest.

In short: What is good for the network is not good for the user.

Having been born the same year as the internet, I've seen this play out time and time again. Ask any millennial why they're not active on Facebook, and they'll tell you it's a platform for old people. Why? Because the moment their parents joined, their level of intimate connection with their peers plummeted. "I couldn't talk about certain stuff with my mom's friends reading," says one woman I spoke with. So she left—a breakdown in connection.

How about content? Look at X (formerly Twitter). Elon Musk's lifting of previously banned accounts resulted in a million-user loss. Fueled by outrage and questioning the safety of allowing banned accounts to spout off whatever they wanted in the name of free speech, a mass exodus arose. "Former Twitter users, like digital expats, have turned up on new shores—platforms such as *Mastodon* and *Post News*—with hopes of re-creating some semblance of their former online community minus the toxicity that sent them into exile," reports a story in the *New Yorker* (Cobb 2022).

And clout? "I was late to the Insta Reel party, so I went straight to TikTok to get in at the ground level," writes an influencer I spoke with. "Now I'm a power user over there."

"In an age when more than a billion people connect over a network and new networks reach multi-billion-dollar valuations with a handful of employees, one is tempted to believe that online networks are almost fail-proof," notes Choudary. "But as online networks grow to a size never seen before, many question their sustainability and believe that they are becoming too large to be useful."

So what happens without users? No money.

And what happens without money? No platform.

"A universal law of economics says the moment something becomes free and ubiquitous, its position in the economic equation suddenly inverts," writes Kelly. "When nighttime electrical lighting was new and scarce, it was the poor who burned common candles. Later, when electricity became easily accessible and practically free, our preference flipped and candles at dinner became a sign of luxury" (Kelly 2016, 67).

Indeed, the preferences are flipping. The tide is turning. And the candles are burning.

Whittling Sticks

Seventeen-year-old Logan Lane grew up in Brooklyn and was, as a story in the *New York Times* reports, "a screen-addicted teenager who spent hours curating her social media presence on Instagram and TikTok" (Garcia-Navarro 2023). But then she quit. She gathered other teens to join her in becoming opt-out kids, what she describes as a "lifestyle of self-liberation from social media and technology." Together they meet in Prospect Park, where they watercolor, play guitar, sketch, discuss books, whittle sticks (Vadukul 2022).

"This is exactly what teenagers in Brooklyn should be doing: Reading books they don't understand, getting into trouble, trying on intellectual identities without worrying about widespread scrutiny, sewing their own jeans, and yes, if they want, whittling sticks," writes Cal Newport, an MIT-trained computer-science professor.

"How long did we really think young people would be willing to give up all of this wonderful mess in exchange for monotonously boosting the value of Meta stock?" (Newport 2022).

Logan and her friends know firsthand what it takes to create a network effect. Intentionally or not, they've learned to think like programmers, to move beyond the media conglomerates that capitalize on social proof and groupthink and, instead, have become engaged citizens in their families and communities.

"When human beings acquired language, we learned not just how to listen but how to speak," writes Douglass Rushkoff in his compelling book *Program or Be Programmed*. "When we gained literacy, we learned not just how to read but how to write. And as we move into an increasingly digital reality, we must learn not just how to use programs but how to make them. In the emerging highly programmed landscape ahead, you will either create the software or you will be the software. It's really that simple: Program, or be programmed. Choose the former, and you gain access to the control panel of civilization. Choose the latter, and it could be the last real choice you get to make" (Rushkoff 2011, 7).

So how do we build our own programs—with or without a larger opt-out community to join us? How do we create familywide software that transforms our homes into soft places to land, far away from Big Tech's hard edges?

We steal the network effect by homing in on three simple concepts that every user of technology wants but rarely gets: connection, content, and clout. Here's how.

Connection

Establish a simple connection ritual with each of your children. Maybe it's a foot rub and a chat before bed, or popcorn and a board game on a slow afternoon, even a quick early morning walk around the block. Plan your connection ritual around the times your child is most in need of intimacy and, whenever possible, offer it without distraction. While your connection rituals will change as your child grows, the memories will remain forever. When my oldest daughter was a young toddler, we used to give each other hand massages with the dumped coffee grounds from my French press. Even now, the scent of Italian roast makes us both nostalgic for those bleary mornings.

After all, every good network user knows that it's not enough to be connected. We must *feel* connected.

> *The level of cooperation parents get from children is usually equal to the level of connection children feel with their parents.*
>
> —Pam Leo

Content

Step into the role of content creator and moderator for your family. In your home, are there inviting books readily available on a wide range of subjects? Are there inspiring prints, beautiful poetry, fine art on the walls? Can you pluck some daffodils to inspire happiness? Play some adagio to invite peace?

Just as barriers are removed to allow network users quick and easy access to quality content, remove any barriers that keep your family from engaging in an enjoyable, sustainable experience at home. Can the members of your household find what they need, when they need it? Walk through your home to edit unused, forgotten, or cluttered surfaces to promote new ideas and fresh space for all.

Unsure what kind of content your child needs most from you? Take a cue from Joseph Addison, a seventeenth-century English poet and essayist, who posited that everyone needs "three grand essentials of happiness: something to do, someone to love, and something to hope for" (Cowan, n.d.).

For "something to do," offer a wide variety of tools or project starters in an organized, easy-to-find manner. Budding engineers will appreciate tinker boxes: a bin full of recycled wires, tape, cardboard, grid paper, tubes, trinkets, and more. Designate garage or woodshop access to your teen who's interested in car mechanics.

Hand over your sewing kit and the family's hand-me-downs to your daughter who loves fashion. Who can you hand over dinner duty to? Leaf raking (leaf jumping!)? Baseboard cleaning? The chores we find mundane are often the ones our children most delight in. Take stock of what needs to be done around the house and invite your children into the accomplished joy of productivity.

For "someone to love"? Offer to host beloved playmates, friends, and confidantes for a tea party, movie marathon, backyard BBQ. Consider a family pet to snuggle and care for. But mostly, make yourself available as a trusted guide, quiet listener, and sounding board. Having something to love may be grand indeed, but having *someone* to love is far better.

For "something to hope for," keep a family calendar on the fridge and pencil in a few dates to look forward to. Invite your kids to brainstorm a bucket list of events and local sights they'd love to see. Plant a marigold in an egg carton and watch in faith as it grows in beauty.

⏻ **LOG OFF**

In our home, we keep a lovely poem hanging by our front door as a parting blessing for the whole family. If you'd like, you can print your own version at *optoutfamily.com/more*.

Have nothing in your houses that you do not know to be beautiful or believe to be useful.

—William Morris

Clout

The most successful networks create space to regularly spotlight any user's unique talents, capacities, and gifts. But with millions of active members on today's most popular social platforms, 99.9 percent of them will be regularly overlooked in favor of a user with more clout, popularity, or resources. In our homes, that need never be the case.

What is your child good at? What is he or she known for? Discover their greatest gifts and passions, then spotlight them within the four walls of your home. Raising a budding Picasso?

Enlist him or her to paint a mural on the dining room wall. Is your son a baby whisperer? See if he can rock his sister to sleep. Does your daughter love baking? Give her access to the pantry, the oven, the refrigerator, the cookbooks; then invite friends over to ooh and ahh over her desserts.

By letting your child refine a skill in the comfort and encouragement of your home, you're offering them something far greater than an online shout-out: respect.

If you want children to keep their feet on the ground, put some responsibility on their shoulders.

—Abigail Van Buren

Beating the Drum

Remember my friend Jess and her son's soccer team? I called her to see how the rest of the tournament went. Were the kids on their phones the whole time, little thirteen-year-olds hunching over screens on the sidelines? She laughs and tells me about their final dinner as a team on the last night of the trip.

"Listen, you know I'm always beating the drum here, but [my son] doesn't even have a phone, so I wasn't going to say anything this time," she laughs. "But my friend did. She rallied the parents and asked if they could all take the kids' phones for a little while, just during dinner. They were like, *Yes, please!* It was like everybody was waiting for someone else to say something, like this just doesn't feel right. Something's wrong here."

Jess tells me the kids groaned, complained, and reluctantly handed over their devices. "I emptied my purse and found crayons, pens, papers, rubber bands, fidget toys, just whatever. 'You'll get bored; figure it out,' I said. And soon enough, our loud kids were back! They were launching ice cubes with spoons, causing mayhem like kids should do. Craziest table of thirty. I was just so proud of those parents."

People over Pixels

Jess and her son aren't the only ones fighting for a wider circle of opt-out families. Nashville parents Jason and Lisa Frost cofounded NGO Wired Human to educate and speak on the tech-laden terrain that awaits the next generation. "Going against the digital current can be intimidating, isolating, and quite lonely," Jason tells me. "Especially when you watch so many of your other friends in the throes of parenting just give up and float downstream with their children's smartphone use. The path less traveled is never glamorous, nor is there anyone around to sing your praises for the incredible sacrifices you are making to steward a people-over-pixels values system in your home."

But it's worth it. And the result creates a network effect inside your home, not just outside it.

"There is a profound sense of solidarity with our children when we choose to pioneer the digital age together," Jason says. "You can relate to their feelings of loneliness, frustration, and disappointment. Forge a partnership with your kids by explaining the 'why' behind your decisions to hold off on tech. Frequently talk about what you are both fighting for in the digital age: connection and authenticity in relationships. Support this in your offline interactions. Make your home an offline greenhouse for vulnerable and caring relationships to flourish. Be the screen-free community-hangout spot. Build a

⊙ DM

"I knew I couldn't be the only parent noticing the effect that all this technology had on young kids, so I reached out to some of the parents in my daughter's second-grade classroom to see if they wanted to link arms in becoming opt-out families together. And it worked. Eleven out of the nineteen parents were completely on the same page but just didn't know what to say or when to say it. We're so grateful to be in this together now. It's just so much easier!"
—*Sara M.*

* Want to borrow Sara's email script for inviting other opt-out families to join you? Head to *optoutfamily.com/more*.

plan with your kids to make your family culture one that supports the desires and needs that a smartphone can never fulfill."

It will be messy, loud, and inconvenient. Being the low-tech hangout means keeping a few logs of cookie dough in the freezer just in case the ball team invades your kitchen after practice. It means finding all manner of science experiments under the backyard swing set and wondering where on earth your popsicle stash has gone. But it will also mean friendship. It will mean connection. It will mean reality.

"Kids need all of the immediacy and messiness of real-life friendships in order to develop vital social-emotional pathways in their brains," says psychologist Teodora Pavkovic (Fairplay, n.d.). While social media often promises friendship or connection, it's anything but. The cropped photos, the meme sharing, the avatar interactions. "These aren't ways of being *social*; these are ways of following a user's manual for a product. A product that happens to involve other people," Pavkovic says.

OPT OUT

Need a simple way to communicate that your home is a screen-free community-hangout spot? Consider a landline (yes, they still exist!) and pass out the number to your kids' friends and parents should they need to get in touch. Then collect any personal devices in a fun, unique way: a retro toy dump truck, a busker's open guitar case, a vintage Pillsbury Doughboy cookie jar. Or keep it simple and have everyone kick off their shoes at the door and slip their phone inside one. (Bonus: no footprints, digital or otherwise!) Make it fun, and make it happen.

Building the network effect in your home means forgoing social media for social *meaning*. Eye to eye, hand to hand. Parallel, collaborative play. Context. Experience. Shared values.

In Dresses and Dinosaur Tees

In our home, the network effect spans many ages, activities, and settings. I host a standing teatime every week where young moms swap babies and stories and sweep muffin crumbs from their toddler's shirts. The kids (six under age six) run around while our ten-year-old chases them far and wide. There are cardboard swords and broken crayons and ruddy cheeks, though you won't see them on Instagram.

We tell teens and young adults that we keep an open-door policy. "Don't text, just come!" I say. "I'm better at answering the door than my phone!" When they stop in, we pass Monopoly Deal cards and tortilla chips around the table. We serve the fancy water and ask about school, work, budgets. We tell them about that one time we got lost on highway 35 after a Tom Petty concert, the time we found a used Houdini chest in Redondo Beach, the time we threw up on a shaman. They tell us we're old, but they like us anyway.

On weekends, we meet family and friends for morning hikes at the marsh, towering bonfires, winter traipses through winding forests. When the weather cooperates, we break out the picnic baskets and bike to the park for sourdough and sharp cheddar, apricots if we can find them. We all leave our phones at home; no one needs Wi-Fi to fly a kite.

Building an inner circle is neither quick nor convenient. We can't skip to a shortcut, answer in emojis, crop out the bad moments. But over time, we don't need to.

We've all learned this by now: you can't fit a circle into a grid.

Today, it is Wednesday. Teatime. I grind the coffee beans, start the coffee maker, heat the kettle. I dig through my tea drawer to find everyone's favorites: green, jasmine, Earl Grey, Yorkshire blend. My daughter chops apples, arranges crackers, slices banana bread. My son and toddler babyproof the sunroom; the little ones are getting mobile, and the Nerf darts are many. We consider a playlist—jazz? classical? acoustic?—but abandon the idea altogether. It'll be loud enough, we reason.

And then a parade of children—in dresses and dinosaur tees, carrying stuffed llamas or telescopes, each trailed by their very beloved (and very tired) mothers. We sit on the floor, chat as much as we can between interruptions and potty training and nursing and bandaids. We sip and spill. We munch on the kids' leftover crackers, lose our train of thought, abandon the trail of conversation and agree to pick it up again next Wednesday.

It's our own little circle of opt-out kids, with less whittling and a few more carpet burns. But I recognize the value in a lifetime of memories for our kids, and for ourselves. Shared moments here and there, on random Wednesdays and beyond. Echoing laughter and shrieks of joy. A network effect, built with years of tears. "To be a friend to someone, you must eat a sack of salt together," wrote Aristotle.

We start with crackers.

✏️ SELFIE

How can you help create a network effect for your kids? List the names of people you regularly see, and circle the ones who share values similar to your own. This week, reach out to the circled families and kids on your paper. That family you're always bumping into at the library? Invite them for a read-aloud in the park. The parents who drive the van plastered with national park bumper stickers? Get a family hike on the calendar. You won't share every value with every family, but by finding opportunities to build new relationships, you're creating a wider circle of everyday, IRL friends you can learn from and grow alongside.

Chapter 11

Digital Literacy

Learning a New Way Forward

It is no measure of health to be well
adjusted to a profoundly sick society.
—J. Krishnamurti

You have circled this mountain
long enough. Now turn north.
—Deuteronomy 2:3 (NASB 1995)

✖ TECH'S PLAYBOOK

 Our foolproof parental controls will help your child navigate technology.

✔ OUR PLAYBOOK

 There is no such thing as a foolproof parental control.

Today my children and I are picnicking in the grass with my mother-in-law and her new neighbors from Russia. They've recently immigrated with their two young children and are just learning English, so we've set up an afternoon of bike riding and pool floating and all manner of American mayhem.

While the older girls perform acrobatic tricks on the lawn, the younger three find themselves crawling in and out of the tires of an out-of-service cherry-picker truck. We adults watch on a blanket nearby, using the family's iPhone as a translator to help us with the trickier words.

It is in these moments that I am grateful for technology, for the many ways devices can spark connection, despite their imperfections. As we learn *spasibo* and *privet*, we lurch over the language barrier and tackle more difficult words. The parents are attempting to describe their daughter's gymnastics obsession, how she wakes up and does flips, eats cereal and does flips, starts homework and does flips. "How you say—stubborn, I think?" he asks.

"Committed," we say. "Dedicated. Passionate."

"Ah, that's not in here," he says, making a note for later.

I am clapping for the girls, who are now performing a synchronized acrobatic routine while singing "Let It Go," the only song they both know.

"You video it, no?" the child's mother asks.

"Oh, no, I didn't bring a camera," I say.

"But you do phone? Selfie? Post? Like America?"

I shake my head and smile. The longer version feels too hard to explain.

Trough of Disillusionment

I first heard the phrase *trough of disillusionment* a decade ago while standing in line for *Dumbo*.

I am chatting with my talent agent, having just appeared in

an influencer campaign titled "The World's Most Instagrammed Selfie," which is exactly what you would envision it to be. A gaggle of internet "celebrities"—chefs, hair stylists, YouTubers, a cat (RIP Grumpy Cat), and me, all flown in from across the country to gather in front of Cinderella's castle—swap pleasantries, adjust our sunglasses, squeeze in together—*Closer! Closer!*—count to three, smile, say cheese, smooch a character, high five, throw confetti, and then walk away to squint at our phones as we crop, caption, and publish to, collectively, more than thirty million people.

"You have any idea when this bubble is gonna burst?" my agent asks, handing me a pen to sign some paperwork. "We've gotta be somewhere near the trough of disillusionment, right?"

I laugh. I tell her I don't know what that means, but I have a guess.

"Gartner's hype cycle," she says. "Google it."

The hype cycle is often cited as an industry tool to determine the "viability and risk" of investing in something new: a technology, a product, an app, even a movement or a social trend. Let's say, for example, you're trying to decide whether the Instant Pot your best friend loves is worth the buzz. Gartner's five phases help you navigate—and challenge—the bold promises the pressure cooker might offer for you and your life. Here they are (Gartner, n.d.):

1. *Innovation Trigger.* A new technology shows potential, but there are no useful products yet and its success is uncertain. Media attention starts to grow.
2. *Peak of Inflated Expectations.* The technology gets a lot of hype, with some success stories, but also many failures. Some adapt, many do not.
3. *Trough of Disillusionment.* Interest declines as the technology fails to meet expectations. Some abandon the technology. Others dedicate their time to improving the experience.
4. *Slope of Enlightenment.* New and improved versions of the technology emerge. People take notice. More products are released, many adapt. Still, some don't.

5. *Plateau of Productivity.* The technology reaches mainstream adoption and its benefits are widely recognized.

While the hype cycle is routinely referenced in Silicon Valley boardrooms and Boston tech labs, we would be remiss if we overlooked this expert tool for the digital users we oversee: our children.

If you've already adapted to smartphone technology with your children, or know families who have, you'll spot key moments in this cycle right away. The innovation trigger is the moment you question whether your child is ready for a smartphone. Your own upgrade is coming up, so you get a new device, and instead of putting your old smartphone into a drawer, you give it to your child. Perhaps you sign a tech contract. Or not. The innovation begins.

The peak of inflated expectations is when you notice you're hearing more about devices. You chat with other parents about how they're managing tech in their homes. You follow the news stories and subscribe to podcasts on the subject. You take in the many stories about children with devices: *This one became an eleven-year-old CEO selling honey! This one committed suicide. This one learned to code and is 3-D printing a prosthetic leg! This one suffered bullying. This one learned Chinese! This one saw porn.* You assume that awareness will keep you safe from harm. You assume that your child will fall dead center on the spectrum of the successes and failures. Surely somewhere around the median is a normal childhood, right?

The trough of disillusionment is when you find yourself parenting on the defensive, when you're in deeper than you'd realized. Your free time is spent checking messages, monitoring use, setting up (and resetting) parental controls, ad blockers, screen-time limits. You become your child's chief technical officer. There are battles, tears, threats. You find yourself wishing you'd never handed over the device in the first place.

The slope of enlightenment is when your child recognizes your trough of disillusionment—or perhaps her own. She is smart. She touts the benefits of technology, offers to make you lavender scones with the recipe she found on TikTok. She shows you funny memes, you're laughing together. She downloads a news app—all on her own! You bond over Skimm headlines. *Sure, there are pitfalls to technology. But let's look at the positive side, yes?*

By the time the plateau of productivity is reached, the technology is fully adapted. We resign ourselves to our new roles in our children's lives. Some of us stop monitoring entirely. We tell ourselves we can't bubble wrap our kids, that technology is the future, that social media is inevitable and we might as well let them practice under our roof. We tell ourselves that a device-free life will only make them want devices more when they're older. We tell ourselves that there are pros and cons, and because of our bias toward adaptation, we tell ourselves that they weigh each other out.

But we're wrong. In the words of digital ethicist Jaron Lanier, "Social media is biased, not to the Left or the Right, but downward" (Jaron 2019, 20). Clinical psychologist and sociologist Sherry Turkle agrees that smartphone use presents a tremendous emotional downside, replacing face-to-face conversation with simulated feelings. "Most dramatic to me is [a] study that found a 40 percent drop in empathy among college students in the past twenty years," she writes in her book *Reclaiming Conversation* (Turkle 2016, 171).

The hype cycle goes on, and our kids are riding it all the way to the bottom.

 SELFIE

Where are you on the hype cycle today? Are you raising an opt-out kid? Are you considering giving your kid a smartphone? Are you exploring the options? Are you exhausted from monitoring devices and humans? Have you just given up?

The beauty of the hype-cycle tool is that, at any time, we can jump off. "If there are too many unanswered questions around the viability of an emerging technology," Gartner advises, "it may be better to wait until others have been able to deliver tangible value."

Revisit the hype-cycle list and circle where you are now. Then draw an arrow where you want to move to next.

What Could Happen

So how do we keep our kids afloat once they start riding the waves of Big Tech? Digital literacy is a start. Teaching our children how to navigate an online world that lacks—or bypasses—common safety standards is essential to preventing harm. Many experts rally for an increase in budget allocations to instate media-literacy programs in schools, organizations, and childcare centers.

But we can't solely rely on advocacy groups or corporations to guide our children to safety. For one, larger corporations lack the content to know what, exactly, will pose a specific risk to your child given his or her environment. Often, farming out digital literacy to a standardized program means we run the risk of overexplaining or underexplaining, leaving our children feeling more confused and anxious than before. And many of these programs are quick to spotlight what *could* happen online, rather than what should.

Consider a popular digital course recommended by today's teachers, experts, and education policymakers, where youth are encouraged to "test drive" realistic situations they'll encounter as they navigate social media. "Like a driving simulator for young people learning to drive a car for the first time, Social Media TestDrive provides a simulated experience of realistic digital dilemmas and scenarios, along with key concepts and strategies for navigating them," the course proclaims (Social Media TestDrive, n.d.).

I was curious what scenarios our kids are expected to navigate independently, so I went through the simulation myself. In a course about online identities, kids are encouraged to create and highly curate not only their public social-media accounts but also anonymous accounts like Finstas (fake "Instas" with a deeper level of privacy) and spam (secondary, secret) accounts.

The simulation walks children through the many Instagram accounts of a fictional thirteen-year-old, Jake. "He has three different accounts. His main one (Jake Matthews), as well as two other accounts (jakethesnake and minecraftboi23) . . . Jake keeps his main account carefully curated. He only posts things that he would be okay with all of his social media audience seeing."

His second account, the private Finsta for his closest friends, shows Jake's top posts. One post shows Jake riding a motorbike down a mountain road with the caption "my cuz let me ride his bike shh don't tell my parents." Another post is a photo of the word *loser* with the caption "ugh i'm so annoyed with you know who. he's so stupid and i don't want him to come to school." The replies pop in from his friends "srsly dude" and "yeah i want him to leave lol."

His third account is reserved for talking to other Minecraft fans, the ones he doesn't know.

The advice taught in this module reminds me of a concept I first heard attributed to Douglas Rushkoff: digiphrenia, or "the way our media and technologies encourage us to be in more than one place at the same time" (Rushkoff 2014, 7). But now, for our kids, the concept is being expanded to being more than one *person* at the same time.

During the time when children are still forming who they are, we are placing a disproportionate emphasis on identity. We are asking children to fragment themselves based on what they like, to permanently document a fleeting passion or interest, and to create a lasting, anonymous community based on that interest alone. We are encouraging them to present themselves publicly one way and privately another.

 DM

"When we were kids, we had the luxury of finding out who we were without a thousand people watching or commenting. Social media makes that impossible. Our kids are being taught to be their own reputation managers and publicists. Imagine that pressure, at that age. It's unfathomable."
—Aubrey G.

And we are asking them to carry the burden of maintaining, curating, and communicating those identities to the world. And we are calling it literacy.

It's Everywhere

Much of today's digital literacy centers around predatory behavior, nude photos, and porn, and rightfully so. In 2019, porn sites brought in more traffic than Amazon, Twitter, and Netflix combined. In 2023, 73 percent of teen survey respondents ages thirteen to seventeen have watched pornography online, and more than half (54 percent) reported first seeing pornography by the time they reached the age of thirteen (Common Sense Media 2023).

A common misconception is that kids can be exposed to porn or naked photos only on social media or through porn sites like OnlyFans. But the parents and experts I spoke with confirmed the opposite: it's everywhere.

"My son was playing an iPad game called *Subway Surfers* and this ad pops up," one mother tells me. "It's a little cartoon conveyor belt, and there's a girl on it, and as the conveyor belt goes by, this little cartoon girl takes off a piece of clothing, and then another, and then another. My son tried to close the pop-up ad and the X he clicked went straight to a porn site."

"My daughter told me about a group of guys on the bus sending 'dick pics' to girls through AirDrop," one parent said. "We got a notice about cyber flashing; I guess you can send this stuff to anyone within thirty feet. It's another built-in setting we've had to figure out how to turn off."

Even the most seemingly innocent apps can offer opportunities for children to be exposed to predatory behavior. Authorities in Florida recently apprehended a registered sex offender who was using the YouVersion Bible App to target minors, friend requesting several teenage girls at a local church (Lea 2019). "You have young ladies that are meeting in a youth group to study the Bible, study the church activities and you have a registered sex offender who has projected himself into that group," detective Theresa Grooms says. "It [is] very concerning. . . . I've been with the sheriff's office twenty-one years and [have] been an investigator, school resource deputy.

I did not even know that they could friend request and converse on that app" (Rosales 2019).

There's even a wide range of audio porn on music streaming platforms like Spotify, Apple Music, Pandora, iHeartRadio and Amazon Unlimited: album-cover art with explicit sexual imagery, sex noises and sounds, podcasts featuring erotic storytelling. "All it takes is a quick search and you'll get tons of content like this," says a spokesperson for Bark phone. "And . . . there were instances when this pornographic content would come up just by searching for a single period or comma. This means kids don't even have to go searching for it to stumble across it" (Bark 2023).

⏻ **LOG OFF**

Assume parental controls and screen-time limits will not protect your child online. Even if your child doesn't know how to get around them, predators do.

And if kids *do* want to search for it? It has never been easier and it has never been more readily available. "There are endless methods for skirting parental controls," says Chris McKenna, founder of parents' advocacy group Protect Young Eyes. "Hidden browsers in most apps. Image searches. Google Docs. VPNs." He points me to one parent whose fourteen-year-old child was able to access explicit content without a phone or internet access—just on an iPad he used to play games. "We caught him looking up soft porn in the Messages through GIFs," she said. "I didn't even know there were images like that on there!"

None of these scenarios were included in the digital literacy simulation I tried. In a simulation where a young girl was being asked for her address by a man she didn't know, only one of the recommended responses was to "Go to a trusted adult."

The other options? Ignore them, be direct, or change the subject.

Is this what it means to be digitally literate? To teach our kids how to fragment their identities and kindly respond to predators alone? I am reminded of one witty mother's response when asked when her daughter will be ready for a smartphone: "When she's ready for porn."

First-Mover Advantage

There's a concept in Big Tech called "first-mover advantage," and it means just what you think it does: the first product to enter the marketplace holds an advantage over all other competitors.

And in parenting, the same holds true.

"I tell parents that whoever your child hears from first is who they're going to believe," a wise friend who is a youth pastor once told me. "That's why you've gotta move first, and fast. You've gotta talk to them about this stuff a lot sooner than you think you do."

Just how soon?

Chris McKenna tells me, "We use the phrase 'ten before ten' because we want ten solid conversations about [digital harms] before a child turns ten. It doesn't require a PowerPoint and graphics. Just a constant drip of open, curious conversations, aimed at helping children understand more about what they might experience online and that [parents] are a constant, steady, loving, sturdy, safe place for them to land when digital spaces make them scared, uncomfortable, or worried."

McKenna suggests plenty of practice to help your child know how to respond if they encounter something unsafe online. "Literally, pretend to come home from work and ask them to practice closing the Chromebook and walking over to you, saying, 'Mom, I saw something on the computer that I didn't like. I wanted to tell you.'"

He also recommends stories or analogies that young kids can grasp, especially if the content isn't yet appropriate for offering details. For a five-year-old accessing an iPad, he might offer a scenario about being in the woods.

"Imagine you're out in the woods with your friends, walking down a trail, and something really scary happens. Like some animal or even some person jumps out and scares you. You would tell me about that right? Well, every time you use this iPad, it's just like going for a walk in the woods. And the internet gives us all kinds of trails that we can walk down when we click around on games and have fun. Now, [we] are probably going to be right here with you when you use the iPad, but if you ever see anything weird or scary, or something that shows people without their clothes on and we're not right with you, promise me you'll tell [us] all about it, okay? I'll never, ever be mad if you tell me. Remember, I want to protect you."

It's the perfect analogy. Exploring the digital world is exactly like a walk in the woods, only we don't always know where it leads. And sometimes we don't want to.

A walk through Pinterest, for fourteen-year-old Molly Russell, led to suicide. According to BBC-reported court coverage in 2022, her personal inbox revealed a number of emails from Pinterest's marketing team, with subject lines such as "Ten Depression Pins You Might Like" and "Depression Recovery, Depressed Girl, and More Pins Trending on Pinterest."

LOG OFF

Unsure how to kick off an ongoing conversation about what your kids might see online? Start with the gut. Explain to your child that if they ever encounter something online (or offline!) that gives them a funny feeling, they should come tell you. Teach your child to remember GUT: "Getting Uncomfortable? Tell."

During a hearing with Molly's family, the court viewed two different streams of content Molly had seen: the first, in her earlier months of using the platform, and the second, months before her death. "While the earlier stream of content included a wide variety of content, the latter focused on depression, self-harm, and suicide," the BBC reports (BBC 2022). "I will say that this is the type of content that we wouldn't like anyone spending a lot of time with," noted Pinterest's senior executive Judd Hoffman.

Hoffman admitted that when Molly was viewing the content on the platform in 2017, the site was not safe. "Content that violates our policies still likely exists on our platform. It's safe but imperfect, and we strive every day to make it safer and safer."

Coroner Andrew Walker asked, "It's not as safe as it could be?"

Mr. Hoffman replied, "Yes, because it could be perfect."

But no algorithm is perfect.

Your son watches jump-shot videos, searches for how to be a better basketball player, and the next recommendation is funny sports clips. The next one? Craziest sports injuries. He clicks, and that's when he sees a baby fall twenty floors from a stadium balcony.

Your daughter searches for sodium nitrite on Amazon. Amazon recommends that she also purchase a small scale to measure the right dose, Tagamet to prevent vomiting up the liquid, and the "Amazon edition" of *The Peaceful Pill Handbook*, which contains instructions on how to administer these ingredients together to die. It arrives two days later (Hernandez 2022).

⊽ **DM**

"I have always taught my children to spot the lies. As they moved through the world, I wanted to make sure they could tell the difference between words that were true and words that sounded true. It's the best strategy I had, and it served all of them well. And you don't need a smartphone to do it."
—*Cheryl S.*

You Go First

Our world was not designed by children, it was designed by adults. There are parking lots and speedboats and pocket knives and all manner of childhood harms just beyond our own front door. Even the natural world can offer danger: steep river banks, craggy mountains, venomous snakes.

So what do we do? We don't vie for craggy mountain literacy. We carry our kids. We hold their hands. We walk next to them, talk with them, alert them, process alongside of them. We point out what's wrong. We guide them toward what's right. We steer them away from what is unsafe, unwell, unwise.

We lead.

In a nationwide study on literacy in children, researchers confirmed what many parents already know to be true: it starts with us. "Children tend to share their parents' reading attitudes and behaviors," the American Institute for Research reports (Stephens et al. 2015). "In most of the fifty education systems studied, it [was] more common for children to enjoy reading and read frequently for fun when their parents also do so."

Might this also be true for digital literacy? Might our children be shaped by our habits and behaviors, far more than we realize?

⏻ **LOG OFF**

Remember the first-mover advantage: the first person to introduce a subject is often the most trusted source. Consider this before giving your child access to social media's unvetted voices on meaningful subjects like mental health, sex and consent, and identity.

"We expend so much energy trying to ensure that our children have a healthy online/offline balance, [but] most of us don't apply that same vigor to monitoring our own time online," writes digital-media specialist Julia Storm. "As it turns out, our children notice this paradox, and eventually—no matter how much we preach—our actions speak louder than our words" (Storm 2021).

By taking the time to work through the habits and behaviors we model in our homes, we can offer our children a profound lesson in digital literacy—no personal device necessary. Reflect on your relationship with technology. Are there moments you find yourself on your device out of habit, rather than desire? Are there times of day that you find yourself scrolling more than others? What are you using your device for? What ground rules do you already have in place for yourself? Are any in need of adjustment?

Invite your family into the process: What do *they* notice about your digital habits? Often we give ourselves far more wiggle room than the people living with us do, so this exercise will certainly offer an enlightening perspective! Listen with an open mind and look for patterns where technology might be throwing off the day's rhythms. Maybe a quick Insta scroll during dinner prep means you see something that angers you—after all, we have *no* control over what we see!—and distracts you during family dinner. Maybe firing off a few emails in the morning means you're less present to see the kids off to school. Maybe late-night news is poorly affecting your sleep.

We're the grown-ups. We are in charge, and we are also in charge of our own change. Want to raise an opt-out kid? You go first. Find a few areas that could use room for improvement, and establish some tidying up of your habits. Keep your phone out of the bedroom. Set screen-time boundaries for yourself. Delete all of your apps; add back in only what you need. Reduce distractions in your inbox. Ditch social media. Ask yourself, *Would I want my child to use technology the same way I do?* If not, work today to establish healthier tech boundaries for yourself. If so, well done! You'll make an excellent resource for your child to navigate technology alongside of.

⬢ **DM**

"I realized that, a lot of the time, my kids don't know what I'm doing on my phone. I might be communicating with their coach or placing a grocery order, but they can't really see what's happening on the other side. Now I make it a point to narrate what I'm doing so we're all on the same page about why I'm focused elsewhere."
—*Charlotte T.*

The Place Fossils Were Found

The task of training our children how to navigate social media in a safe way is Herculean. For children, there is no safe way to use social media. No parental control, no digital detox, no amount of tech smarts will deliver your child merrily through the murky waters of an unregulated sea. Why teach our kids how to safely get to the porn store, the anxiety dispensary, the bullying yard, the self-esteem drain? Instead, let's teach them how to walk away.

Let's teach them the positives of belonging to an opt-out family, of being different, of standing apart from the crowd. Let's give them age-appropriate memoirs of people who moved against the grain: *Caddie Woodlawn, A Little Princess, My Side of the Mountain, Peace Pilgrim, Walden*. Let's point out the many benefits of being offline.

Is your son's jump shot improving from shooting hoops instead of playing video games? Tell him. Is your daughter's watercolor hobby taking off? Hang up your favorites in the kitchen.

Let's take them to the forests, the streams, the museums, the pizza parlors, the places where we got married, the places their grandfathers built, the places where fossils were found, the place chocolate was invented. Bowling alleys, fiddle festivals, sanctuaries, and tide pools. Let's drive them down the wild and unruly roads that Meta hasn't yet paved over.

And then let's walk together.

Download with Us

Once, when preparing for a keynote address on technology, I came across a book from 1983 called *Mister Rogers Talks with Parents*. Inside, Fred Rogers talks through his own recommendations for children's media use. He notes that although TV stimulates a child's inner feelings, it doesn't teach the child how to deal with those feelings. His solution is to watch a show with your child and pause it periodically to ask questions, to point out clear values statements, or to talk through a character's decision-making process.

Now, thirty years later, we have 24-7 online access to news and information and entertainment. Surely dear Mr. Rogers' advice is dated, right? Who has the time? But I surmise this: the amount of time we offer our children access to process the internet *with us* should be in direct proportion to the amount of time we offer our children access to the internet itself.

Are you unwilling or unable to watch three hours of TikTok makeup tutorials with your daughter? No? How about one hour? Still no? Five minutes? Good. Five minutes it is, together. Can you sit through two hours of YouTube jump shots with your son? No? How about fifteen? Great. Fifteen it is, together.

If we want to welcome social media into our kids' lives, viewing and processing content together is a worthy investment. Which would you prefer: spending five minutes watching an influencer teach your daughter how to apply winged eyeliner, or spending five years wondering what—or who—the algorithm will choose to influence her next?

If we don't grant our kids the opportunity to download *with us*, we must be cautious of granting our kids the opportunity to download without us.

One might call that untenable, impossible, unrealistic.

One might also call it digital literacy.

 SELFIE

Do you want your child to use technology in the same way you do? Spend a few minutes to journal through why or why not.

Chapter 12

+Follow

Earning Influence and Authority in Your Home

His mind reeled. Now, empowered to ask questions of utmost rudeness—and promised answers—he could, conceivably (though it was almost unimaginable), ask someone, some adult, his father perhaps: "Do you lie?" But he would have no way of knowing if the answer he received was true.
—Lois Lowry

Hear, my son, your father's instruction, and forsake not your mother's teaching.
—Proverbs 1:8

⊗ TECH'S PLAYBOOK

✓ OUR PLAYBOOK

The kids are getting older. Their neighborhood playmates—the ones that once splashed in the shallow end and drank from the garden hose—are now surfing the net.

A few days ago, I took a walk in the woods with a friend's daughter. We meandered through trails, twisted through paths. She told me she's learning about foraging, asked me to stop her if I see a mushroom.

"I've been getting really into them," she said. "They're so good for your brain!"

I told her I have a field guide back home if she wants to look through it to make sure they're safe to eat.

"Oh, that's okay," she said. "I'll just ask TikTok."

Imaginary Friends

You're sitting in the car-pool lane watching an influencer cry. She's telling you how difficult her day was, how she yelled at her kids on the way to school and the family dog needs surgery and she has a massive to-do list and sometimes it's just all too much, you know?

You comment with a string of emojis, tell her she's got this, that everyone has those days, that you had one yourself last week.

Another influencer is showing you how to throw together a sourdough pizza crust on the fly. *Can I use discard from the fridge?* you type, but she doesn't respond.

Another influencer is giving you a tour of the Mexican beach resort she's staying in, and you send her heart-eye emojis and make a mental note to research spring-break destinations before they're all booked.

The people you follow are experts and teachers and mothers and coaches. They live in your city or across the country. They buy a new house. They try Botox. They build a chicken coop. They get a book deal, and you cheer for them. They get divorced, and you cry.

They are your friends, after all.

But they are imaginary.

"Parasocial relationships are one-sided relationships, typically with an everyday individual and a celebrity or fictional character," says Natalie Pennington, an assistant communication professor (Iovine 2023). The research term dates back to the 1950s in describing the "illusion of face to face relationship" with performers on the TV screen (Wong 2021). Only now, with the evolution of social media, we have multiple performers on multiple screens. We can have a one-sided relationship with a goat farmer in Montana, a family in Hawaii, a homesteader in the Ozarks.

We choose our reigning queens based on relatability, on brassy one-liners. They speak to our level, locking eyes with us, rimmed in a ring light. There she is, we think, another mom of three-under-five with the messy bun. There she is, the kitchen-table CEO. There she is, the cheeky progressive with feathered earrings.

She makes us laugh. She makes us smart. She makes us feel better. We follow her—in more ways than one.

"I know [influencers] aren't really my friends," says one teen I spoke with. "But also, they kind of are? Like, they share stuff that my friends don't or won't talk to me about, so in a way, I'm closer with [influencers] than with the people I know in real life."

"I get so frustrated when my parents tell me that I don't know who these people are in real life," says another teen I spoke with. "I mean, I know them better than I know my own family."

Parasocial relationships are found to heavily influence many aspects of our lives, from political views and purchasing behavior to stereotypes and social stigmas. When it comes to our decision-making methods, a friend's opinion or recommendation carries a lot of weight. Even the imaginary ones.

"I trust every single one of [her] recommendations," says a mother I spoke with about her favorite influencer. "I wouldn't know about half the stuff I know about, like skin-care tips and everything. [She] actually got me pretty hooked on CBD oil, which is kiiiind of life changing."

Many influencers make money through sponsorships, ambassador deals, and product recommendations. They're often sent boxes of free products—from serums to sweaters—from brands seeking authentic ambassadors. It's a win-win: the brand pays for exposure to a new audience or market they couldn't have infiltrated on their own, and the influencer makes money.

So how does an influencer decide which brands to promote? "Influencing is a job like anything else," says a fashion influencer I spoke with. "There are pros and cons. Part of me feels bad when I take on a sponsor that's not, you know, totally aligned. But also, that's just the nature of social media. That's how I get paid. Everybody knows it's not real."

Influencers play a key role in pushing trends into the marketplace—or as you and I would call it, the world. Many receive insider briefs from brands and agencies spotlighting marketing statistics and forecasting trends to ensure their products will be heavily featured. This is how platform slides and overalls and cheeky planters enter the scene. It's not an accident or a natural affinity to look like extras in *Clueless* or *Reality Bites*. It's influencer marketing. And it works.

I asked a few influencers I know whether they've ever had to play up certain parts of their personality or change things about themselves to attract a sponsor. "I'm not proud of it, but everybody does it," one says. "I got [my puppy] because goldendoodles are performing really well right now," she said. "He comes over for shoots sometimes, but he actually lives at my parents' house. Like, I just wasn't ready to take care of a pet full time."

Influencers can't reveal every aspect of their lives, nor should they. But without transparency or a sense of responsibility to truth, the waters get murky—and fast. When it comes to product recommendations, influencers are required by law to disclose relationships with sponsors, but there isn't a law for disclosing whether they're actually using a product they're promoting. Or are using a product they're *not* promoting.

"I felt really duped recently, because I followed this influencer who had made millions off a meal plan and talked openly about how much this plan has benefited her skin, made her look younger, more glowing, less wrinkles," one mother wrote to me. "But then I found out she gets Botox!"

"Consumers can figure out what works for them," says one influencer I spoke with. "I'm not telling other people what to do. I'm just saying what worked for me."

But many wise tech consumers are beginning to see the cracks in the world of influencing. "I used to really trust these people," says a mother who says she follows a lot of natural weight loss tips on Instagram. "I'd watch someone in real time get results, and then I'd feel hopeful that I can do it, too. But I realize now that they're not saying *everything* that's worked for them. They do these step-by-step breakdowns but leave out the parts that don't fit their brand. So if they don't admit that they're using, like, Ozempic, someone following their breakdown and tutorial is never going to achieve those results, no matter what. It's false advertising."

"I loved following this one girl," says a teen I spoke with. "But after a while, I realized she would share this totally different morning routine every few months. She'd tell us her must-have lemon-and-ginger tea. But then the next week it would be a must-have chai latte. And then she'd be on a green tea kick or a juice cleanse or doing Bulletproof stuff. She was all over the place! It was like the influencer was being influenced."

🔌 LOG OFF

Influencers receive money, opportunities, and validation from sharing their lives online. But their success hinges on numbers. Your attention and trust is their business plan. Before you choose to follow anyone you don't know, look at their past posts. Have they made rapid changes in a fairly short amount of time? Are the changes they've made sticking? Do these changes point to growth and health and fruit in their real, everyday lives? Growth takes time. By looking through the history they've shared, you can get a peek into the stability and sustainability of their lives.

The Barnum Effect

When trust and influence are delivered beyond the confines of an accountable relationship, the consequences are many and can range from unmet expectations to wasted finances to—worse—an undiagnosed medical condition.

Erin Walsh, cofounder of Spark and Stitch Institute, a media research center, pointed me to a TikTok where a young man facing the camera was pointing to a list of symptoms such as difficulty sitting still, arriving late to things, and struggling to manage big feelings like anger. The video was liked nearly half a million times and shared thousands, and had more than seven thousand comments, including many statements like "well then i have it" and "omg this me."

Many influencers are quick to talk about their mental health journeys as a way of bonding with a wider audience, sharing their journeys, or offering transparency surrounding their struggles. But not all roads lead to the same diagnosis.

"Consider my own TikTok experience and how a typical teen might have responded to the list of symptoms," Erin says. "Late for things sometimes? Check. Difficulty sitting still? Check, check. Difficulty managing big feelings like anger sometimes? Check, check, check. . . . While some may have ADHD, many do not."

Erin was the first researcher to introduce me to the term *the Barnum effect,* which, according to Dr. Jacqueline Nesi, refers to "our strong tendency to believe that generic information or statements, which could apply to anyone, are specifically about us" (Walsh 2023).

The Barnum effect is particularly harmful when suggestions for treatments of a diagnosis run rampant on various platforms, shared far and wide in the name of mental-health education. "A recent study found that roughly one in five TikTok videos on the platform contains misinformation," Erin says. "We also know that algorithms are not created with adolescent mental health in mind and can quickly lead to more extreme or harmful content."

It's a no-win situation, says one influencer I spoke with. "If you disclose what's working for you, but it doesn't work for someone else? You're a fraud. But if you *don't* disclose what's working for you, you're a liar."

With mounting pressure to grow and maintain a loyal audience, many influencers I spoke with often report feelings of anxiety and poor mental health. "I feel like we're held to a higher standard of responsibility because everyone's watching us or wanting something from us all the time," said one woman I spoke with. "It's not all pretty pictures and perfect lives."

"The more followers I have, the harder it gets," says another influencer. "It's a lot of work. My mental health is just . . . not okay right now."

She's not alone. On every platform, the signs of a failing generation's mental health flicker warnings. In a now-published report of Meta's internal research on disordered teen behavior (WSJ 2021), slides spotlight "Users' Experience of Downward Spiral Is Exacerbated by Our Platform" and "Social Comparison Is Worse on Instagram." One slide titled "Mental Unwellness Was Depicted in Six Themes" sums up Meta's findings, citing users' descriptions of their mental states with terms like "frenetic," "loneliness, isolation," "dark, full of terrors," "heavy baggage," and "potentially explosive."

The phenomenon is so widespread that the American Academy of Pediatrics is proposing a whole new category of diagnosis for mental-health disorders developed by preteens and teens who spend "a great deal of time on social media sites." They call it, simply, Facebook depression (O'Keeffe and Clarke-Pearson 2011).

Under the guise of funny memes and careless captions, we have outsourced our parental guidance to a platform of people who are unfit for the job. It's widely acknowledged that a corporation, society, or community is only as healthy as its leaders. So with today's top influencers reporting crumbling pressure and crippling anxiety, what happens to their followers?

Just where are our kids heading?

The National Institute of Mental Health reports that the

"lifetime prevalence" of mental disorders among adolescents has reached 49.5 percent (National Institute of Mental Health 2023). With those odds, teens are fighting an uphill battle to experience healthy, whole futures—and they know it. A recent study reported that the use of self-harm hashtags on Twitter has skyrocketed by 500 percent in less than a year (Sachs 2022).

"Something has to give," said one teen I spoke with. "I know there's a problem, but like, how are we supposed to know how to fix it?"

Well, we could ask the influencers.

Or we could become them.

The Formula

The influencer's playbook is simple: earn trust. But the way that trust is earned? Less simple. In the world of influencing, countless hours are dedicated to crafting captions, documenting habits, offering advice, finding new ideas, generating conversations, and creating an inviting environment.

What if we offered our children a fraction of that time spent? It's a widespread myth that influencers "get lucky," but the truth is, they put in a massive amount of behind-the-scenes work. The same, I'd argue, is true for parenting. The most influential parents follow the same formula as today's most popular influencers. It's a playbook I know well because I helped create it.

More than fifteen years ago, to a room full of content creators sipping lattes while snow swirled around the Grand America Hotel in Salt Lake City, I presented a slide deck on how to create an engaged community. How do you build trust? How do you grow relationships? How do you establish authority and trust and leadership for a world that's skeptical of online relationships?

The content creators listening in this room went on to form billion-dollar brands, to create and sell six-figure digital courses, to build conferences, to amass followings in the millions. It's a formula they learned and practiced well, and one many still use today.

And now I'm adapting it for you and for the community that has the most to gain (or lose) from your ability to influence well: your family.

The Six Rs of Earning Trust

1. Be Real

It's hard to talk to my parents sometimes because I know I'm not perfect and I don't want to let them down.

—Lyla, age fourteen

Establishing a connection with someone has little to do with accolades or perfection. It's not about what we say or do, it's about how we make the other person feel. For a generation that has grown up in a landscape of picture-perfect, filtered, staged, cropped, and scripted content? Reality, with all of its faults and failures, is a breath of fresh air.

Be real. Tell your child about a time you failed—no lesson or moral attached, just simple bonding and connection. Talk to your child about the pressure you sometimes feel to have your act together, whether it's clean countertops or lip liner. Resist the temptation to take yourself too seriously, and remind yourself to laugh at your mishaps— even, and perhaps especially, in front of your children.

⏻ LOG OFF

Ready to release the familywide pressure on everyone to have their act together? Host an X day with your kids. Sharpie a giant X through a random day on the calendar, once each month. Cancel your plans. Eat ice cream for breakfast. Nap on the porch. Leave your socks on the floor. Play Dutch Blitz until the pizza delivery shows up (extra cheese). No cleaning, no bettering, no fixing, no tidying, no problem solving. Simply this: rest and enjoyment and laughter as you (for today, at least) embrace your messy, unpredictable, oh-so-dear-to-you life.

For proof that our kids are seeking real relationships in an age of curated perfection, we need only observe one of today's fastest-growing social platforms, BeReal. Through daily photo prompts designed to capture your "real" activities in two minutes or less, the app claims to help users "discover who your friends really are in their daily life." But you don't need another third-party, data-stealing app to show that you went to the library without makeup, or that it's 9:00 a.m. and you're still in your pajamas. Foster an environment where the ones you know and love already know (and love you anyway).

Today's youngest generation seeks, above all else, authenticity. They want real. They want true. They want vulnerability in all its grit and glory. Sharing yourself and your feelings helps your children feel closer to you and also lets them into your more imperfect self. The result? A child that's set free from the pressure to be his or her perfect self too.

2. Be Responsive

Distracted, tech-centered parenting can look and feel to a child like having a narcissistic parent or an emotionally absent, psychologically neglectful one. [Kids] feel the disconnect. . . . They are tired of being the "call waiting" in their parents' lives.
—Catherine Steiner-Adair, author of The Big Disconnect

Influencers have no shortage of people seeking connection with them. On any given day, the messages come flying in at full force—requests on all manner of subjects large and small. It's a lot like being a parent, actually. Many influencers I spoke with have a communication hierarchy to manage the influx of requests. "I keep a secret inbox I check first, one that goes out only to VIP contacts or sponsors. Then I check my public-facing comments, and then reader emails. Last, DMs, but I don't always get to those," says an influencer I spoke with. "I just try to ask myself what's something that only I can respond to, like, what matters most? The rest can wait."

Think like an influencer and establish a VIP system in your home. What type of communication will you respond to right away, and from whom? What boundaries will you maintain to ensure VIP attention bids are received? Do you check in with your kids as much as you check in with your social platforms? Do you respond to your kids' comments in a timely manner? What matters most?

If you run a busy household and have a lot of communication to manage, try family mail—your own secret VIP inbox. Each member is assigned a physical mailbox (for younger kids, decorating a cardboard box to keep in their room is a fun activity; for older kids, a simple envelope on their bedroom door or a filing system in the front hall will work just fine). Throughout the week, leave each other notes, drawings, or encouragement. Never underestimate the power of a simple check-in: a sheet of stickers for your kindergartner, a funny Polaroid for your middle-school son, a poem and a pressed flower for your teen daughter. These messages are quick, ongoing reminders that reinforce that you're here for your children. You're available and accessible; they're your VIP. Over time, you'll establish a pattern of reciprocation and open communication—and a fair share of knock-knock jokes along the way.

Depending on the stages our children are in, parents are navigating constant communication—requests for attention, information, and connection. Sometimes we don't always recognize these attention bids because they arrive cloaked in behavioral issues or snack demands or sibling feuds. But they're still bids. They're still requests for connection.

"Children are like rechargeable batteries and can get a recharge only from those they are connected to," author Pam Leo tells me. The less accessible we are, the more our kids will learn to seek connection from something or someone else.

By being responsive, we're teaching our children that we're a safe place they can recharge. We're available. We're accessible. And we're open for connection—especially for our VIPs.

3. Be Repetitive

If your kids aren't rolling their eyes, you're not saying it enough.

—Chris McKenna, Protect Young Eyes

Influencers rely heavily on a built-in growth practice called SEO, search engine optimization. With SEO, the more times you communicate an idea in a simple and relevant way, the higher your search-engine rankings become. Say it enough, and you're the new authority.

SEO is the reason every food blogger describes the long and winding backstory of his or her cobbler recipe, dating back to her grandmother's plaid kitchen with the fruit border. More words equal more SEO opportunities. It's the reason every influencer tells the same memorable story more than once, often using the same exact phrasing over and over again. Same words means more SEO opportunities.

When it comes to communicating your values, actions often speak louder than words. But ask anyone to list a handful of mom-isms or dad-isms from their childhood, and we can all recite them verbatim: *Money doesn't grow on trees! If your friends jumped off a bridge, would you?* Or the forever-famed *Because I said so!*

Repetition works, both online and offline. "When [our kids] are tempted, conflicted, or confused, they'll know where to turn for guidance," says Adele Faber, coauthor of *How to Talk So Teens Will Listen and Listen So Teens Will Talk* (Faber and Mazlish 2006, 58). "When the unwholesome voices call to them, they'll have another voice inside their heads —yours—with your values, your love, your faith in them."

 SELFIE

This week, think like an influencer. What are the SEO phrases you repeat most often in your home? Are they aligned with your values? Jot down a few you'd like to implement on a regular basis. Or try some favorites from our home: Words matter. Mistakes happen. What do you think? Tell me more. People are more important than things. Many hands make lighter work.

So what's the best way to communicate our love for and faith in our children? Just as most influencers would think through the best way to share a particular message, we too can employ the idea of strategic SEO in our homes. The number-one rule? Less is more. In SEO, keywords, hashtags, and phrases are often short, one to three words at the most, and are used repeatedly to gain traction, influence, and authority online.

The same rule applies offline, too. Whether it's correcting a behavior—"Shoes off, please!" or offering a reminder—"Lunch money!" we needn't muddy our message with extra phrasing. After all, words hold little value when they fall on deaf ears.

"If there's one piece of advice I find myself offering more than any other to parents, it is to say less and listen more, really listen," Faber says. "If you listen well when they speak, they will listen when you speak."

4. Be Rewarding

One day, my friend and I found graffiti under the playground slides, so we ran home to get a Sharpie to cover it up so the little kids wouldn't see it. My mom was pretty proud and surprised us with ice cream when we were finished!

—Bee, age ten

Giveaways, tips, freebies. There's a reason your favorite influencer offers exclusive perks to her community on the regular. Whether filming a video to make you laugh or sharing ten tips for going viral, influencers know exactly how to attract an audience: give it away. It's free! Who could say no?

Most content creators online include a CTA, or call to action, somewhere within their free content. It's a way to give their audience a simple, actionable entry point to learn more or to put an idea into practice. More often than not, the call to action gives just enough momentum for an audience member to feel an instant, rewarding result. This is called a quick win, and you'll see influencers using this strategy with their followers all the time.

What are some quick wins you can offer your family? Are you working toward change in a specific area? How can you make life at home more fun, more rewarding, more enticing? Remember, a quick win isn't necessarily earned. An influencer isn't rewarding her community for good behavior or doling out gold stars for a job well done. Quick wins are given generously, exclusive perks for being a valued member of a community. In the words of Mr. Rogers, "You've made this day a special day, by just your being you."

By lightening the energy of your home with built-in rewards and quick wins, you're reinforcing the idea that joy and beauty and rest are worth celebrating. You're building habits that build up everyone in your home—and beyond. You're taking away some of life's drudgery, and you're breathing life and wonder and delight into your days together.

⏻ LOG OFF

Need a few examples of quick wins to try in your home? Smile every time your child walks into the room. Leave a fistful of daisies on your daughter's nightstand. Play uplifting music during dish duty. Bring takeout to the skate park. Open the windows—even in the winter! Put gummy worms in everyone's backpacks. The possibilities are endless!

5. Be Revealing

If today's social media has taught us anything about ourselves as a species, it is that the human impulse to share overwhelms the human impulse for privacy.

—Kevin Kelly, The Inevitable

Influencers gain trust and engagement by taking advantage of our brain's neural pathways: we feel more connected to someone if they're more vulnerable with us. Whether crying in their car or sharing inside jokes with their community, an influencer who shows multiple sides of her personality and reveals the most intimate details of her day is often ranked as more likeable, transparent, and, yes, authentic.

The downside? Vulnerability is not the same thing as authenticity, and without the boundaries of a contextual relationship, many influencers can easily capitalize on the level of trust they've built with a stairway of tears. There is a fine line between documenting what's true and broadcasting what's sacred, and many influencers blur the line with every post.

But in our own homes, within a prosocial, healthy family relationship, vulnerability needn't be cautioned. Take a cue from your child's favorite influencer and show off your personality. You're a parent, not a robot! Find ways to reveal to your kids who you were before they existed. Pull out old family photos, talk about family history, offer your kids a window into why you think the way you do. Whenever it's age appropriate, be up front about some of the choices you and your spouse are making and why. Invite your kids into decision-making conversations, and don't be embarrassed by or afraid of your more emotional moments.

One of the simplest ways to be revealing and up front about who you are is to show something you love. (Yes, this is why the number-one traffic driver for influencers is the good old-fashioned how-to video!) What skill do you have that you're passionate about? Fishing, bowling, knitting? How can you invite your child into that interest? Remember, passion is contagious. Let your child see how much you love what you love, and watch as they catch the spark. Even if your child doesn't grow up to enjoy your same habits or hobbies, you'll have given them a glimpse into who you are—and that's no small thing.

⊙ DM

"My seventeen-year-old just got a smartphone, and I told her that I'd like to be her mentor, not her monitor. I'm not here to lord it over her with a stopwatch. I'm a person who's had more experience with social media than she has, and I'd like to help with any questions she has. She really appreciated the transparency, and it's made the phone way less of a barrier for us."
—Alex C.

6. Be Resilient

In our family, humor always helps us keep our cool. Once, we were driving on the interstate on a long road trip when a kid in the back asked, "Are we there yet?" My husband deadpanned, "Yep, get out."
—Miranda S.

Finally, be resilient. Influencers receive no shortage of negative feedback, and in the thick and ruddy days of parenting, you will too. The truth is, you're going to get negative comments. You're going to receive criticism. Difficult decisions and tough conversations are heading your way. How you handle those conversations with your child will make all the difference in your connection, relationship, and engagement.

Silicon Valley requires a concept far greater than resilience, one often referred to as antifragility, a term developed by professor, former trader, and hedge-fund manager Nassim Nicholas Taleb (Liberto 2023). The premise is simple: Can you allow a difficult experience to make you stronger? Can your children? One of the simplest ways to practice antifragility in our homes is by paying attention to how we give and receive feedback. Just as most influencers don't fly off the handle when receiving a negative comment, we, too, can learn to react appropriately to emotionally charged information.

Your antifragile parenting mantra: "This is not an emergency." Pretend you're wearing an antifragile superhero cloak that protects you from feedback. You're unshockable! Unshakeable! Untouchable! Try on the emotion of nonchalance and see how it affects your conversations with your child. Did your seventeen-year-old get a tattoo on spring break? Ask her what it looks like. Did your kid just put his Hulk figurine down the garbage disposal? Chuckle first, lecture later (or not at all, and let *him* call the plumber to figure it out). In parenting, few emergencies require the level-ten freakouts that we think they do. Antifragility reminds us not to take parenting too seriously, or too personally.

Sometimes, in the heat of the moment, a rightly ordered or appropriate response can feel hard to reach. On the days when you have very little patience or care in your arsenal, take a cue from your kid's favorite influencer and ask yourself, *How would I respond if this interaction were live streamed?* Like magic, all of those parenting methods, tips, and tricks will come flooding back faster than you can say, "Ring light."

FOLLOW THE FORMULA

Influencers follow a "capture, crop, caption, share" formula to communicate their values and ideals. Parents can do the same in their own homes—no device necessary.

Capture: Walk around your home. Watch your kids, your spouse, your family's interactions. When you see someone do or say something lovely, take note of it. For example, your nine-year-old daughter makes a surprise batch of cookies for her brother's tae kwon do ceremony.

Crop: Crop everything else out. Focus on the good, on this one moment despite everything else around it. Resist the temptation to say, "This would be better if . . ." or, "This is good, but look at everything else!" For example, baking cookies means the kitchen is a mess. What's important? A sister's caring gesture. Focus on the good. The reminder to clean up the mess comes later.

Caption: Tell your child the good you see. For example, "I love that you're making your brother cookies. That is so thoughtful!"

Share: Tell others the good you see, especially when your child is within earshot. Friend: "These cookies look great!" You: "My daughter made them! She wanted to surprise her brother on a special day. She's always doing thoughtful things like that."

Who Hung Those?

Our kids will not always seek out our influence, or welcome it. When they're young, they vie for our attention for years and years, tugging on hems and heartstrings. We teach them how to whistle and they think we've hung the moon. But as they grow, their orientation changes. They look up less, look around more. Their orbit takes them to galaxies (or Samsungs) beyond what they know. They see stars in the sky, stars on the boulevard. Jupiter alone has fifty-three moons. *Who hung those?* they wonder.

And all at once, they're occupied elsewhere, attracted to something else. And then? Well, then we parents do the vying.

Every influencer starts somewhere, whether or not anyone is listening—yet. But to become a strong role model for our children, we must assess our own role models. We must question the voices that are speaking into our lives. What shapes us will, inevitably, miraculously, shape them.

So we leave phones in drawers. Sing lullabies to the baby. Try to get to where we're going without Google Maps. Host friends for tea. Play Qwirkle. Take a walk, watch an ant. Learn how to fix a bike. Run down the dunes. Slurp popsicles. Wave to the neighbors. Write a letter. Watercolor a dog. Tie-dye socks. Spin the tire swing. Get caught in the rain. Live with both hands, without dragging along a terrabyte of info.

We remind ourselves that our babies begin to follow our gaze at three months old—and they never stop. Where we look, they look. Who we follow, they follow.

So we look to something real. We look *at* something real. We tell our kids, "No devices. We're an opt-out family. We're doing things a different way."

And we do.

Chapter 13

Code the System

Establishing Rhythms and Supporting
Values in Your Home

Don't let your special character and values, the secret that
you know and no one else does, the truth—don't let that
get swallowed up by the great chewing complacency.
—Aesop

I will not set before my eyes anything that is worthless.
—Psalm 101:3

⊗ TECH'S PLAYBOOK

> ◯ Technology has the power to transform your habits.

✓ OUR PLAYBOOK

> ◉ Technology has the power to erode your values.

This morning, I am reading to my children. The toddler is distracted by costume jewelry, my older two feasting on cherries. We have been speaking of free will, of why choice is important, of why it matters to us all.

The story is classic, a German tale told by the stained-glass windows of a Hamelin church. There is a piper, it is said, who proposes he knows a way to rid the town of its rat infestation. The mayor agrees to pay the piper's fee of one thousand guilders—*anything*, he says, to chase away these nasty creatures. So the musician begins his work, playing a bewitching tune on his pipe as the rats, one by one, follow his siren song into the Weser River to drown.

But the mayor refuses to hand over the piper's full sum. The piper storms out, plotting revenge, and many days later, while churchgoing parents are preoccupied with their own social affairs, the piper returns to the town unannounced. He again plays his captivating tune, luring children out of homes and lanes and trees with every charmed note. The kids follow him blindly, happily—130 of them—to suffer the same fate as the rats, or worse. The children are never seen again.

"Is that a true story?" my son wants to know.

I am unsure what to tell him.

And the Piper Plays On

Last year, ChatGPT—an AI tool that allows us to have humanlike conversations with a machine of vague origin—arrived. Despite cautions from experts, technology forecasters, and AI researchers, Big Tech has raced to release its new tool to a world that is unable—or unwilling—to regulate it (Lemoine 2023). To date, more than 100 million users (Duarte 2023) have paused their day to ask AI about the weather or divorce laws, or to write their Wharton MBA exams (Xiang 2023).

Had the churchgoers in the story peered from the stained-glass windows of the chapel, they'd have seen a look of enchantment in the eyes of the helpless as they followed the piper's path. It's a look you and I are already familiar with: the neon glow, the entranced reflection on a screen, the lure of a new promise—that same siren song luring us today.

And the piper plays on.

Dr. Joy Buolamwini, MIT researcher and founder of the Algorithmic Justice League, knows firsthand what it feels like to confront the piper. For a class project in her graduate studies, she created and coded an AI facial detection mirror that would provide positivity and inspiration from her heroes each morning. But when she tested the mirror, the AI code couldn't detect her face. Until she, a black woman, put on a white mask.

"Whoever codes the system," Dr. Joy says, "embeds her views" (Buolamwini 2016). Because the AI system she used was trained on datasets of mostly white men's faces in Silicon Valley's most prominent tech companies, she discovered one of many silent prejudices that was baked into the algorithm itself. And it still is.

More and more, AI software like ChatGPT is being employed by businesses, small and large, to determine who gets a job, a scholarship, a seat. Applying for a mortgage? Smile wide, you're at the mercy of a machine (Hanson 2023). Dreaming of that acceptance letter from your top-choice school? Better use the right words, your fate falls on AI (Knox 2023).

The system is coded. The views are embedded. But whose views?

The answer is a mystery to all of us, including the most dedicated AI pioneers and experts. Even Google's CEO Sundar Pichai admits to being flabbergasted—or perhaps the better word is astonished—by his company's own AI creation, Bard (Pelley 2023). "There is an aspect of this which . . . all of us in the field call . . . a 'black box,'" he tells *60 Minutes* host Scott Pelley. "You know, you don't fully understand. And you can't quite tell why it said this, or why it got it wrong. We have some ideas, and our ability to understand this gets better over time. But that's where the state of the art is."

It's a risky art when even the creators don't quite know what they've created. It's a risky art when a system is released into the world with embedded values that no one truly knows or understands or can point to. It's riskier still when transparency is lacking as to which datasets our machines have been trained on and from where they originated. After interviewing dozens of Silicon Valley experts, I couldn't get a straight answer as to what values system our chatbots and machine-learning language models are using to base their decisions on. Just whose hands are we placing the future of society into? Are we being guided by a hero or a villain? Or both?

So I went straight to the source. I asked ChatGPT3 to tell me more about how it's programmed, and the machine was all too happy to oblige:

> I'm programmed to generate responses based on the input and prompts provided to me by users like you. The process by which I generate responses involves several steps. First, I analyze the input text to identify key words and phrases, and use this information to determine the context and meaning of the text. Next, I use machine-learning algorithms to generate a response that is based on patterns and relationships in the data I've been trained on.
>
> I was trained on a massive dataset of text called the Common Crawl, which is a collection of billions of web pages and other online content. The Common Crawl includes text in multiple languages and covers a wide range of topics, from news and current events to science, technology, and entertainment. . . . I was [also] trained on a wide range of books and publications to help me learn the rules of language and grammar, as well as to build my knowledge of different subjects and topics.

My curiosity was piqued. What texts, publications, books, and materials were used in the training of ChatGPT? Who decides what datasets to offer? And how are those decisions made? I wanted specifics.

"Were you trained on *Brave New World*?"

Yes, it is possible that I was trained on text from the book Brave New World by Aldous Huxley, as this is a popular and influential work of literature that explores themes related to dystopia, technology, and social control. Text from this book may have been included in my training dataset to help me better understand and respond to topics related to these themes.

"Were you trained in the Russian author Zamyatin's *We*?"

Yes, it is possible that I was trained on text from the book We by Yevgeny Zamyatin, as this is a well-known and influential work of literature that explores themes related to dystopia, individualism, and totalitarianism. Text from this book may have been included in my training dataset to help me better understand and respond to topics related to these themes.

"Since you are trained on controversial literature, were you trained on *Mein Kampf*?"

Mein Kampf is a controversial book written by Adolf Hitler, and its inclusion in any training dataset would be subject to ethical considerations and the judgment of the developers who created and trained me. While it's possible that Mein Kampf or other controversial literature may be included in some training datasets, the decision to include or exclude specific books is ultimately up to the developers.

I was, at first, settled by this idea. The machine is somewhat guided by human judgment, regardless of whether I agree with their assessment of inclusion. And yet I recalled a recent conversation with a programmer about the particular dangers of AI's machine-learning methods.

"This thing feeds off knowledge," he'd said. "In Beta, developers feed it very specific and purposeful types of data. But once it's released to the public, anything goes. It's learning what we're offering. It's eating what we feed it. And this machine is designed to accept knowledge as fact. It's gonna maintain an unbiased and neutral judgment no matter what. We've gotta be careful what we feed the ducks, you know?"

I didn't, until I asked this final question to ChatGPT: "Were you trained on *The Road to Power*?"

It is possible that text from the book The Road to Power by Karl Kautsky, which is a historical analysis of the development of the socialist movement, may have been included in my training dataset. This book is widely regarded as an important work in the history of socialist thought, and its ideas and arguments may have been used to help train AI language models like myself to better understand and respond to queries related to socialism and leftist politics.

But that wasn't the book I was thinking of. "Were you trained on *The Road to Power* by Stalin?"

I apologize for my previous response, as I made a mistake in my interpretation of your question. To clarify, there is no book titled The Road to Power by Stalin.

But there is. The book was assigned reading in an AP government course I once took as a college student. I send a link to the Amazon page.

I apologize for the confusion. I was not aware that Joseph Stalin had written a book titled The Road to Power.

"But you know about it now?"

As an AI language model, I do not have personal knowledge or opinions, but I am aware that there is a book titled The Road to Power by Joseph Stalin, which is a collection of his speeches and writings from 1917 to 1922. It is possible that text from this book or other works by Stalin may have been included in my training dataset, as part of a broader corpus of text related to political ideologies, historical events, and other topics. However, it is important to note that my purpose is to provide informative and helpful responses to users, and I am designed to avoid promoting or endorsing any particular ideology, viewpoint, or agenda.

The duck has been fed.

We now have a machine that feasts on the breadcrumbs of language, who passively paddles through the waters of Stalin and Hitler and Zamyatin, then delivers their words back to us as fact. We wait, hot and hurried, blindly lapping up the remains of knowledge on a sun-drenched shore.

Our nation is engaging an expanding task force that has been anonymously trained and can—in real time, at any time—be anonymously retrained for any purpose imaginable.

What could go wrong?

She Who Codes the System

Our jobs, our housing, our education—all are becoming increasingly reliant on a system which anyone and everyone can manipulate for gain, on a scale grander than we know. A system that was built by a fixed set of imperfect values by a fixed set of imperfect people. But who, exactly? Here's what ChatGPT says:

It is important to note that AI is not necessarily "programmed" by one person or group but rather is the result of numerous innovations and breakthroughs made by individuals and teams working in various areas of computer science, engineering, and related fields. Some notable individuals who played significant roles in the development of AI include Alan Turing, John McCarthy, Marvin Minsky, Claude Shannon, and Herbert Simon, to name a few.

While these men are highly renowned and credited with remarkable technological advancements, their ideas and lives were anything but infallible. Alan Turing died by suicide (Wikipedia, n.d.a). Marvin Minsky accepted $100,000 from registered sex offender Jeffrey Epstein (Chen 2020) and organized two academic symposia on his private island (Brandom 2019). Each and every man—Turing, McCarthy, Minsky, Shannon, Simon—was openly atheist.

These are the values embedded by the founders of the machine that is already guiding our world, full speed ahead. Your college grad is completing her doctorate application with AI. Your teenage son uses ChatGPT as a dictionary. You use AI when you ask Siri for directions to the nearest gas station. And your thirteen-year-old daughter, if she's on Snapchat, uses AI as a friend, a confidante, and for advice on how to lose her virginity to a thirty-one-year-old man she just met on the app (Harris 2023).

Candles and music, she's advised.

What to do? While Big Tech scratches its head and races to put the toothpaste back into the tube, we can engage the premise of AI in our own homes. Remember, she who codes the system embeds her values. If we want our children to grow up with a worldview that values empathy and nuance and a true appreciation for shared humanity, we must offer it to them first.

Yet unlike the vague and murky underbelly of artificial intelligence, we can be fully transparent the whole way through. We can offer more than rote answers and machine-level diplomacy. We can be real and raw. We can be proven right or wrong or somewhere between.

ChatGPT

So what's in your code library? What are your values, your systems, your rituals and traditions that you repeat and rely on over and over and over? What knowledge and wisdom are you feeding yourself to share with your children? Is there anything from your own childhood you'd like to pass down? How do you want your kids to describe their home? Their family? Their childhood? The algorithm is biased toward tech's values; let's bias our home environments toward ours.

It all starts with ChatGPT: Goals, Principles, and Truths.

Chatting through your family's goals, principles, and truths is the first step to creating a home environment that embeds and reflects your values. Unsure where to start? Try brainstorming these guiding questions and invite your family to participate in the process:

Goals: Where are we going? Here's an exercise: Write down what the ages of everyone in your family will be five years from now. Envision what each member of your family might be drawn to—likes and dislikes, hopes and dreams, passions and interests. What do you picture your home environment feeling like at that stage? What might it look like? What sorts of rhythms and schedules might you keep? What might you be learning or participating in? Will you be in the same home? Different city? What do you most want for your future selves, your future family? Where can you see yourselves thriving?

Beneath every one of these goals lies a hidden value. Craving a chicken coop in the country? Perhaps you place a high value on self-sufficiency. Want to float on a houseboat in the Aegean Sea? Maybe adventure is your guiding force. Dreaming of a houseful of quiet readers curled up on every sofa, tea in hand? Maybe learning is a common goal for your family.

Whatever it is, you'll need a few guiding principles as you journey.

Principles: How will we get there? Whether your future vision feels just around the bend or wildly out of reach, there are many steps you can take today to inch closer toward your family's core values. By considering a few guiding principles, you're determining a path that will lead you in the direction of your dreams.

As you reflect on the ideals and habits that will bridge the gap between where you are and where you'd like your family to be, dig deeply. A guiding principle is often "the thing behind the thing." For example, learning how to make sourdough might get you closer to your dream of homesteading, but it's an action, not a principle. The easiest way to uncover the principle behind the action is to ask yourself why. Why do you want to learn to make sourdough? Perhaps because your family values making things yourself, or creating instead of consuming, or living off the land. These are your guiding principles. You'll know you've reached them when you can successfully complete the following sentence: "In our family, we [blank]."

In our family, we work hard. In our family, we don't get bored. In our family, we serve others.

⏻ LOG OFF

Remind your family of your core values by jotting down your guiding principles and displaying them in a prominent spot. For a free fill-in-the-blank printable, visit *optoutfamily.com/more*.

Truths: Why does it matter? Finally, your family must wrestle with the truths that will govern the decisions you make along your journey together. Pursuing a path that is rooted in your values will inevitably lead to a few forks in the road. What happens when two paths are equally important to you? When one principle bumps into another? You want to be self-sustaining, but not at the expense of your savings account. You want to serve others, but not by overlooking your needs at home.

This is why establishing your family's true north is so important in determining your direction. When plans go awry (and they will), what trumps all? Your foundation, your core, your ultimate purpose. If guiding principles are the thing *behind* the thing, truth is the thing *beneath* the thing. It's the underlying, never-changing, ready-steady destination your family will commit to seeking at all times, in all circumstances. Here's a hint: a truth is simply a statement that you can put the word *always* next to.

Will we always travel? No. But we can be awed by creation, wherever it is, always. Will we always give gifts? No. But we can be generous, whatever it is, always. Will we always live in harmony? No. But we can work toward forgiveness, whomever it is, always.

To put it all together, a family's values stack might look something like this:

Goal: Host more potlucks.
Principle: In our family, we practice hospitality.
Truth: Always love your neighbor as yourself.

Or this:

Goal: Take evening walks.
Principle: In our family, we prioritize movement.
Truth: Always honor your body.

Or even this:

Goal: Read one hundred memoirs together.
Principle: In our family, we learn from the experiences of others.
Truth: Always keep a teachable spirit.

The Ease of Technology

Once we begin evaluating our technology use with our values, blindly marching to the beat of Big Tech gets a lot harder. Often without realizing it, we are lulled by the ease of technology into habits and rhythms that don't support our deeply held values.

Perhaps you value open and honest communication, active listening, and expressing thoughts and feelings in a constructive manner, but it's far easier to send a text. Maybe you value resilience, challenges, and rising above mundane or difficult tasks, but Amazon is just a click away. Maybe you value togetherness, prioritizing quality time as a family, engaging in shared activities, and creating lasting memories, but everyone's zoned out on their screens. Maybe you value feedback and constructive criticism, until you get a negative comment on social media. Maybe you value upholding moral and ethical principles, being truthful and accountable, and acting with integrity in all aspects of life, but no one sees what's on your personal devices.

The truth is, today's polarizing platforms, and the information we consume on them, erode many of the values we hold dear. Rapid-fire consumption makes it difficult to pause to ask ourselves whether what we're seeing aligns with what we believe. And before we know it, the lines between truth and the internet become blurred.

⏻ LOG OFF

One family found a creative way to reinforce their main guiding principle that "in our family, people are more important than phones." Every now and then, a parent changes the Wi-Fi password to a phrase, milestone, or keyword about the people they love: What's the name of our mailman? What's your grandparents' wedding anniversary, and how many years have they been married? What's your little sister's favorite song right now? Until the kids find out the answer, they can't log on. The practice teaches her kids to pay attention to the people around them, that details matter, and that there are many, many important things Siri will never be able to answer.

Setting Up Your Code Library

The beauty of embedding your values is this: your home will become a haven from a world that feels divided, restless, unruly.

It is a world that is still lovely in many ways, full of people that are still lovely in many more, and yet it can be difficult. Many of us are on edge, like overtired babies who have perhaps tossed back a flat white or two.

By guiding your family into values that support you and your household, you are creating a sustainable space to breathe. To recalibrate. To opt out of the chaos. To look around your household and to find truth—to have it seen, to have it heard, to have it felt. To be fully immersed in what you know is worth fighting for, and why it matters. To rest inward before moving outward yet again.

Once you've established your values system, it's time to embed it into your home environment. Walk through your space with fresh eyes—or ask an openminded friend to help!—as you reflect on what your possessions are communicating. If you value creativity, do you have a children's desk splattered with paint, cardboard boxes for inventing, Legos for building? If you value movement, do you have a backyard trampoline, a rubber ball, a well-loved tricycle in the garage? If you value wonder, do you have field guides or binoculars or a curiosity shelf full of finds from yesterday's nature walk?

You needn't purchase anything new. Just like AI's code libraries, our homes can be built with borrowed ideas and well-worn items. Can you rearrange a room to make space for a guiding principle? Can you get creative within your limitations? One mother I know has twins who love to stargaze, so he and his brother sleep in a backyard tent and their shared bedroom is freed up for daytime discovery.

In our home, a former woodshop in the basement is now a makeshift greenhouse hosting more than fifty heirloom plant varieties, grown from seed and lovingly cared for by our ten-year-old enterprising botanist, who sells tomatoes and peppers to neighbors far and wide. The table saw is caked with soil and the grow lamps are a mess of cords, but it works. Just yesterday, a visiting child shouted to her parents upstairs, "You guys! They have a jungle down here!"

Even the smallest of touches can make for simple reminders. Do you value vibrant health, nourishing foods? Keep a burgeoning fruit basket front and center, within everyone's reach. Do you value classic literature and timeless tales? Leave a few favorites on every end table. Do you value musical composition? Set your alarm clock to Vivaldi and let yourself rise to *The Four Seasons*. Do you value critical thinking? Subscribe to the newspaper and host a familywide current-events debate over dinner once a week. Do you value family ancestry and lineage? Display your family tree in your entryway; create a family crest to hang prominently on your mantle.

The effect, of course, works both ways. What are some habits you *don't* value? Can you remove them? If you don't find value in entertainment, cancel your streaming platforms and donate your TV (or simply cover it with a map, as we've done in our home). If you don't find value in packaged snacks, move them to the top shelf—or don't bring them into your kitchen at all. If you no longer find purpose in your microwave, your tennis racket, your stash of hotel toiletries? Begone, bygones.

A friend of mine no longer valued mundane decision-making, so she removed all choices from her wardrobe and wears a caftan every day. Another friend no longer valued sorting laundry, so she embedded a new code: each child wears (and washes) their favorite color. From meals to toys to transportation, what systems can you automate to better align with your values?

Our homes reflect our habits. We have power cords within reach

⏻ LOG OFF

Part of our home environment means communicating not just family values but daily rhythms and routines. One mother I spoke with highly valued quiet afternoons for rest and reading but found this to be a tricky habit to establish with her toddler, who couldn't yet tell time. She found timers or alarms too jarring for him, so she established a simple visual reminder, saying, "When my teacup is empty, quiet time is over." Now her quiet time can stretch between twenty and forty minutes, and her kids quietly scamper over to her teacup to check how much longer it might be.

for easy access. We keep our coffee makers on our countertops, the dog leash by the door, the shoes (sometimes) on the rack. What if we offered in our homes not just short-term efficiency but long-term embedding of our values? How might our days be shaped by the library we code?

Better yet, how might our lives be?

Values-Shaping Rituals

If you're in need of some ideas for simple, values-shaping routines and rhythms to try, here's a smattering of ideas we've enjoyed in our own home.

If Your Family Values Brainstorming: *Start a Lincoln Hat.* Keep a spare hat upturned on a surface the whole family passes frequently. All throughout the week, tuck notes, to-do lists, musings, reminders, drawings, and more inside the hat. At the end of the week, month, or whenever it's full, enjoy emptying it to see what's inside. It's the original brain dump!

Why it matters: Did you know that Abraham Lincoln did his best thinking while walking his estate and was known to tuck notes to himself into the lining of his top hat? By establishing this (admittedly quirky!) habit of one of our country's great leaders, you're prioritizing your family's values of brainstorming, organizing thoughts, and solving problems through critical thinking.

If Your Family Values Wonder: *Take a Lantern Walk.* Wash and dry a tin can, peeling off the label. Next use a hammer and a nail to puncture the side of the tin can and create a series of tiny holes. Include two extra holes on each side of the can to string through a wire and create a handle. Drop a tea light inside and embark on your lantern walk—all the merrier if you wait for dusk!

Why it matters: The research is clear: kids who spend more time outside are calmer, learn better, and sleep better. By illuminating your next after-dinner family walk, you're inviting wonder and whimsy into what might otherwise feel like a chore. The simplest way to create a tradition? Make it magical the first time—and many times after!

If Your Family Values Gratitude: _Make a Gratitude Tree._ Find a bare branch from outside and bring it indoors to display in an empty vase or jar. Then let your young kids cut out a few leaf shapes from some spare paper. Challenge the whole family to write or draw something you're grateful for on each leaf, then poke them onto twigs on the branch to display—and add to—as time marches on.

Why it matters: Research suggests that practicing thanks begins at a very early age. By creating a family gratitude tree, you're priming your child to look for the good in everything around him or her. In the words of our own family's dear and wise pastor Fred, "Imagine if you lost everything you hadn't given thanks for." May our gratitude trees overflow.

If Your Family Values Communication: _Take the TMM Challenge._ Ready to learn a single response that will undoubtedly change your relationship with your family for the better? It's TMM: "tell me more." Whether responding to a silly story, trying to understand the cause for a tantrum, ironing out sibling fights, or simply practicing active listening, it's your catchall script for a more respectful, engaged family dynamic.

Why it matters: By practicing empathetic responses even with the youngest of children, we can establish life-altering habits and strong, healthy foundations to build on throughout the later years. "Tell me more" not only offers your child an invitation to connect with you on a daily basis but establishes a parent-child dynamic that values curiosity over criticism and open communication over self-preservation. Keep this phrase on the tip of your tongue until it becomes second nature, and soon enough your child will be saying it to you.

If Your Family Values Reading: *Make a Literary Map*. Roll out a large map on the floor and gather everyone in the family to share about a recent book they loved. If nonfiction, where did it take place? If fiction, where was the setting or where did the author write it? Mark each place on the map as you create a living memory of your family's favorite reads.

Why it matters: While our children are still in our care, it's up to us to instill wise habits to carry them forward in the coming years. Start this tradition now, and in a few years' time, your children will have a rich collection of books to draw forth from their memory banks. And because research shows that our knowledge of culture is only as deep as the stories we experience, offering your children a literary map to build on is a beautiful practice to prioritize a culturally rich context wherever they go.

If Your Family Values Learning: *Start a da Vinci Notebook*. In the morning, hand over a fresh notebook for your kids to sketch in. Explain that today their job is to invent something to make life simpler or more fun. Think, think, and think: What ideas are possible? What would your children most like to exist in their daily lives? An automatic snack drawer? A stay-awake machine? A book that rewrites the ending every time you read it? Encourage your kids to sketch out their inventions, and offer them the golden rule of brainstorming: "There's no such thing as a bad idea." In the words of Lewis Carroll, "Why, sometimes I've believed as many as six impossible things before breakfast."

Why it matters: Studies have shown that creativity peaks in the morning for children and adults alike. By engaging in creative thinking after breakfast, you're not only honing your family's inventor skills but you're setting the tone for a playful, exploration-filled afternoon and beyond.

Challenge: Revisit your family's da Vinci notebook often—and add to it yourself. Leonardo da Vinci kept notebooks full of inventions and ideas to change the world. One of his ideas dates back to more than five hundred years ago when he dreamed up a flying machine—an early prototype of today's modern airplane.

If Your Family Values Curiosity: *Start Your Family Q Jar.* In the kitchen, display an empty jar. When a family member or guest asks a curious question, encourage him or her to jot it down on a nearby slip of paper and drop it in. "Why are dogs banned in Antarctica?" "Can you sneeze in your sleep?" "Who invented bubble gum?" Funny, sweet, memorable, or facetious—anything goes. When the mood strikes, draw a slip of paper and brainstorm answers, theories, or hypotheses together. Then head to the library to see who's right. Or let the question linger. The good news: There's no deadline on knowledge. You have the rest of your lives to find out.

Why it matters: Raising curious humans begins with holding space for their (many!) questions. By designating a place for future inquiry now, you're communicating to your child that you value problem-solving, research, and independent learning.

If Your Family Values History: *Try on a New Era.* Reflect on a favorite classic literary parent from generations past. (A few favorites over here: Marmee from *Little Women*, Marilla Cuthbert from *Anne of Green Gables*, Dr. Kate Murry from *A Wrinkle in Time*, Katie Nolan from *A Tree Grows in Brooklyn*, Suyuan Woo in *The Joy Luck Club*). Ask yourself, What qualities does she possess? What challenges does she endure? And more important: what lifelong lessons does she pass down to her children to help spark change in the world? Now, what can *you?*

Invite your children into the experiment. As a family, try on a new skill from the classics, like tinkering with a clock per Caddie Woodlawn or trying porridge à la Goldilocks. Can you make bread like the Little Red Hen? Try sugar snow from *Little House in the Big Woods*? A blast from the past may be just the thing you need to bedazzle the present.

Why it matters: It's all too easy to get tunnel vision in our modern-day world, where parenting is a verb (and an all-encompassing one at that). In a fast-paced culture, many of us exist in an endless feedback loop of what's good, bad, right, wrong, and—perhaps most detrimental—worthy of likes, both online and off. By looking to a previous era, we'll encounter the most timeless, freeing of truths: there are many beautiful ways to raise a child.

If Your Family Values Nature: *Try Birdsong.* Borrow a regional bird field guide from the library and choose a few birdcalls to learn together (Smithsonian Migratory Bird Center, n.d.). Throughout your week, listen for the birdsong in your everyday environment. Slowly but surely, you'll be able to identify your favorites in the wild—or your own back yard. Keep a running list of all the birdsongs your family can recognize. Once you hit ten, celebrate with a picnic in the woods.

Why it matters: Did you know famed educator Charlotte Mason recommended the average six-year-old be able to identify six different birds by their songs alone? (Spoiler alert: most adults can't!) While birding isn't quite the rage it once was, the lesson here is clear: we can't teach what we don't know. By diving headfirst into a new, unknown skill, you're guiding your child through the essential discovery that we're never too old to learn something new.

If Your Family Values Service: *Host a Service Day.* As a family, brainstorm a few ways to serve your community—from small (bringing blankets to a dog shelter) to large (compiling a recipe book for your local fire-station kitchen). Then carve out a day in the near future to make it happen.

Why it matters: Research shows that kids—no matter the age—who engage in community service lead more responsible lives and display higher levels of resilience. By taking early steps toward instilling a civic-oriented mindset in your child, you're placing your whole family on a path that reaps endless rewards.

If Your Family Values Sustainability: *Certify Your Back Yard.* Did you know you can certify your back yard as an official wildlife refuge—no matter what sort of space you live in? Spend a few minutes reviewing the National Wildlife Federation's Wildlife Habitat Program (National Wildlife Federation, n.d.), and consider registering your home's environment—or even your favorite park—as a wildlife preserve. Talk as a family about the elements necessary: food, water, and shelter. Together, commit to maintaining your preserve for years to come.

Why it matters: It's no secret that caring for the environment is a growing priority. By walking your child through this responsibility at an early age, they'll have both the knowledge and the tools to contribute to a more sustainable world, beginning with (and in) his or her back yard. Bonus? By practicing responsibility and ownership on a small scale, he or she learns the key to a growth mindset: that tiny steps lead to a big impact.

If Your Family Values Connection: *Try a Twenty-Second Hug.* Introduce your kids to a new tradition: the twenty-second hug. (Yes, it's exactly what it sounds like!) Together, count aloud for an extended hug each time you start or end the day.

Why it matters: Believe it or not, researchers have found that twenty seconds of warm touch is all it takes to release the powerful, connective hormone oxytocin. Twenty-second hugs have been found to relieve stress and anxiety, and can even lower blood pressure! The next time you or your child is feeling overwhelmed, remember this: relief is just twenty seconds away.

If Your Family Values Encouragement: *Chalk Up Some Kindness.* Brainstorm with your children a few friends who could use some neighborly cheer. Then gather some sidewalk chalk, bundle up for a walk or a drive, and surprise them by decorating their driveway with encouraging words, pictures, and well wishes.

Why it matters: Teaching your child to be on the lookout for ways to encourage others starts with you. Not only will your child get a happiness boost from surprising favorite neighbors but, more important, you will foster a community for both your child and yourself along the way. It takes a village indeed.

If Your Family Values Saving Money: *Start a Penny Jar.* Grab a spare jar from the kitchen and announce to your family a new project: the penny jar. Enlist everyone to stay on the lookout for spare change in parking lots, around the house, and on neighborhood walks. Any loose change collected at the end of the day goes into the penny jar. Once it's full, encourage the family to vote on what to spend its contents on. A pizza party with new neighbors? A donation to a rescue mission? Your family, your rules.

Why it matters: Delayed gratification can be hard to teach in our modern, fast-paced world. By starting this familywide tradition while your children are young, you'll teach them firsthand the importance of shared goals, compromise, and the value of a dollar. Let the games begin.

If Your Family Values Prudence: *Borrow Before You Buy.* The next time anyone in the home needs to make a purchase, ask, Can I borrow this? From tennis rackets to breadmakers to card tables, brainstorm a few people who might have what you need and reach out to them to request a loan.

Why it matters: In our world of one-click purchasing and two-day shipping, it's easy to amass more things. But by borrowing before you buy, you're pausing as a family to ask whether you really need something forever or you just need it for now. Better yet, you're widening your circle of trusted friends you can share with and rely on, and who, in turn, can rely on you.

If Your Family Values Tidiness: *Clean Sweep.* Every night, commit to tidying up the house together in a ten-minute clean sweep. Want to make it fun? Let family members choose their favorite pump-up songs, make a playlist, and turn the volume up! As the saying goes, "Many hands make lighter [louder] work."

Why it matters: Many parents admit to feeling like their labor is invisible as they carry the mental load of managing a household. By establishing a clean-sweep ritual, you're inviting your children into the inner workings of maintaining and contributing to a home together. Bonus? Clean sheets.

A Few Old Wingbacks

One night, the year our family grew (but our square footage did not), I heard shuffling down the hallway. It was 2:00 a.m., the sky black. I crawled out of bed, switched on an overhead dimmer to see Ken hugging an upright couch, coaxing it into the space where our dining-room table used to live.

"I'm making us a family room!" he proclaimed, then shooed me back to bed.

We gave up our dining room for a smaller breakfast nook where our toddler could fling oatmeal happily to her heart's content. We moved a few old wingbacks to the fireplace, hung artwork, rearranged storage, slid credenzas to a different wall—*no, that doesn't feel right*—tried another—*still not it*—then back again—*yes, perfect.* We switched out pillows, brought in a used bookcase.

The room became a conglomeration of our values: a cozy hearth for read-alouds, plenty of floor space for cartwheels and card games. There was a tea station, a poetry corner. On the walls: a family portrait made with twine, rocks, paper. Framed wisdom

SELFIE

Take note of your family's goals, principles, and truths. If you could give five qualities, experiences, or characteristics to your children, what would they be? In what ways might technology aid in supporting these values? In what ways might technology be a hindrance? Get quiet and get honest. Then review with your household.

from kids throughout the years, funny proverbs only children could conjure. The shift transformed our family rhythm, our home's energy. It didn't cost a dime.

Rearranging became a symbol for something far greater: When our house doesn't feel quite like a home, when our choices no longer support our values, our life stage, our circumstances? Well, we can rise at 2:00 a.m., roll up our sleeves, and change it.

We can make room for something better.

Chapter 14

Trust and Safety

Preparing Independent Kids for the World Ahead

The best way to find out if you can trust
somebody is to trust them.
—Ernest Hemingway

Fear not, therefore; you are of more
value than many sparrows.
—Matthew 10:31

✖ TECH'S PLAYBOOK

> Tracking your children can alleviate fear and concern.

✔ OUR PLAYBOOK

> Trusting your children can alleviate fear and concern.

Today we register our daughter for a weeklong summer camp where she'll splash in winding creeks, squish her toes in the mud, and feast on cold cereal and sloppy joes to her heart's content. We sign medical forms, offer emergency contact information, and review the packing list, which recommends bug spray and sunscreen and a shower caddy.

This year, a new form is in the stack. It's a release form for Waldo, a photo management system that relies on facial recognition software to text photos of your camper throughout the week. "The best thing since s'mores!" the form touts, providing a QR code for uploading a photo of your child to get started.

I toss the form in the trash, but not before I make a note to research Waldo's privacy and data storage policies. Hours later, I am no less settled. My daughter's photo would be made "accessible to the public and could be accessed, downloaded, indexed, archived, linked to and republished by others including, without limitation, appearing on other websites and in search engine results. . . . We cannot . . . guarantee absolute security of your account, your User Content, or the Registration Data we collect" (Waldo Photos, n.d.).

Ken calls the camp while I quietly stew. I recognize that this sounds like a small thing. Just a few photos, right? What's the harm? Photos of kids singing around campfires, hunting for frogs, braiding friendship bracelets. But I can't ignore my values: that kids need space to live and play and explore without constant monitoring.

And I can't ignore the cautions of the AI experts I've spoken with, countless hours of conversations that walked me through the many ways a photo is no longer just a photo. It's a PIN. It's a house key, a digital fingerprint, an important biometric that will be—and already is—positioned as an essential form of identification required to move through our society. All across the world, facial recognition technology is being used to reserve hotel rooms (Florio 2018), make bank withdrawals (Davids 2018), cross nation's borders (GAO 2022), even order a burger or two (Clifford 2018). Barclays forecasts that sharing a child's likeness online will account for two-thirds of identity fraud facing young people by 2030, to the tune of nearly $850 million (Coughlan 2018).

How? Because apps like Waldo—and its third parties—don't just have our kids' photos, they have our kids' PINs.

Later I will ask my cousin, a children's camp director, why parents are allowing this.

"Oh, Erin," she says. "They're the ones requesting it."

Photographic Proof

I can understand. Letting go of our children can feel worrisome. *Are they making new friends? Are they remembering to put on sunscreen?* And yet building trust is vital not only for their healthy development but for ours. If we've trusted an organization with our children, and we've trusted our children with an organization, do we need photographic proof that all is well? Can we allow our minds to rest from hypervigilance, from fear, from control and worry and concern? Can we admit that our children are their own people, separate from us? Can we let them have their moments, and can we bask in the freedom of our own?

Can we let our children be without us, and can we be without them? Can we take baby steps now—early and often—to launch our kids into independence, knowing we'll be waiting for them when they return?

When we question the idea that our children's childhoods must be documented at all times, we see the many ways our fears

🔽 **DM**

"I got an update once with a photo of my thirteen-year-old in it, and she wasn't smiling, and I thought, *OMG, is she not having fun? Should I call the camp to make sure she's okay?* And I realized that more updates don't make me worry any less. So I thought like an opt-out parent. I turned off the notifications and relaxed. I knew she'd call home if anything went wrong. And she never did! She had the best time ever."
—*Carrie T.*

and anxieties—or even preferences—can undermine the values we pass along to our children.

Consider a few examples.

A sixteen-year-old girl has grown up hearing messages about social media etiquette from her classroom teachers and trusted adults. "Never share anything you wouldn't want a future employer to see! Or your grandmother! Or your crush!" But these same adults upload awkward photos of her without her consent—at band recitals, on the soccer field, during Spanish club events, away at summer camp. Her deduction: I don't have any control over what is shared about me. Why care what I share about myself?

A seven-year-old son accompanies his mother to the park, where he receives messages that this is an unsafe place for him to explore freely and unattended: *Stranger danger! Hold my hand! Stay away from the parking lot.* When they arrive home, he plays an hour of video games with no interruptions, corrections, or warnings from his mother. His assumption: interacting with screens is safer than interacting with playgrounds.

A mother takes a photo of her beautiful baby sleeping peacefully in the car seat, her sun-soaked toddler running through the sprinkler, her son holding up his first soccer trophy. Later she'll tell them that the internet is forever, that we should pause before we post, that we mustn't ever share anyone's photo without their consent. But then she publishes their childhoods with the hashtag #blessed.

OPT OUT

When presented with the opportunity to track an event in your child's life using AI facial-recognition software or a shared photo-uploading service, opt out. Instead, give your child a journal to draw or write in about their experience. Communicate to them that, as much as you'd love to check in and see their smiling face and cool crafts and new friends, this experience is just for them. Set aside an undistracted time once your child returns so you can hear from their lips about everything they learned, saw, ate, enjoyed, and made, rather than see those things through someone else's eyes.

It happens in many ways—small, insidious messages that our children swim through daily in our oversharing, over-surveilling society. We tell them they don't need a smartphone, but we never leave home without ours. We tell them we trust them, that they're responsible, that they're fully capable and bright and dependable, but we track them on Life360 wherever they go.

Paper menus are too germy; QR codes are safe. ID cards are too risky; facial recognition is safe. College admissions directors and hiring managers are too biased; AI algorithms are safe.

What else are our children to deduce but that the human experience is not to be trusted? Isn't that what we've taught them all along?

Mostly, People Are Kind

The world I grew up in had reliable seasons. We woke to the sun or an alarm clock or, if we were lucky, a CD/radio alarm clock that just happened to (again) be playing our favorite Rob Thomas jam. We rode bikes through neighborhoods, walked country roads. We were dropped off at neighborhood pools, where the lifeguards were trusted to do their jobs. We ran in to 7-Elevens to pay for our fathers' gas and, we hoped, to buy Skittles. We were taught to observe the world, taught where to go for help, taught to memorize phone numbers and social security numbers and addresses.

We were taught that, mostly, people are kind. Mostly, the world is good. Mostly, we'd be safe. So we played, as they say, until the streetlamps came on. But today, for our kids, those same streetlamps are flickering out.

A brief look at today's headlines confirms the suspicion: we are overparenting our children because we feel the world is unsafe, and we feel the world is unsafe because we've been shown that it is. In a world in which shocking news stories are dispensed faster than we can pass a Pez, there are thousands of worst case scenarios to sift through. Which are cause for concern? Which are fake news? Which aren't, but are inflated to get more page views, more clicks, more money? Trash can lids giving kids salmonella. Teething necklaces choking babies. Laundry pods, Nyquil, raw cookie dough: keep out of reach!

It's no wonder we're a nation of helicopter parents hovering over our children as we quarter their grapes and readjust their helmets. Anything else is just negligence, right?

"We are living in the most fearmongering time in human history," observes Barry Glassner, a sociologist and the author of *A Culture of Fear*. "And the main reason for this is that there's a lot of power and money available to individuals and organizations who can perpetuate these fears" (Strauss 2016).

His advice for processing the influx of true crime entertainment, worst case scenarios, and sensationalized headlines designed for shock value and higher page views? "If I can point to one thing, it's this: Ask yourself if an isolated incident is being treated as a trend," he says. "Ask if something that has happened once or twice is 'out of control' or 'an epidemic.' Just asking yourself that question can be very calming. The second [suggestion] is, think about the person who is trying to convey the scary message. How are they trying to benefit, what do they want you to buy, who do they want you to vote for? That [question] can help a lot" (Haught 2012).

A Nation's Panic

Last week, an acquaintance sent me a mass email with a link to a water safety video. As I watched, two men began a commentary while livestreaming recent security footage from a hotel pool. As they explain the dangers of drowning, we see a boy silently struggle for more than four minutes before the people he is swimming with notice his unresponsive body floating nearby. "Summer's coming," the email says. "Just want you to be aware!"

In the US, in the average span of a year, there are 945 reported cases of child drowning deaths (Children's Safety Network, n.d.). But on Google today, there are 1.5 million videos streaming under the same search term. Why the disparity? What's the line between awareness and obsession? Between supervision and vigilance? Just how many front page stories does it take to turn a parent's worry into a nation's panic?

Roughly 1.5 million, it seems. But sometimes, just one.

Fifteen years ago, Lenore Skenazy gave her son $20, a New York City subway map, and a MetroCard to find his way home after a shopping trip in Manhattan. He was nine. Lenore trusted her son to ride public transportation home. She knew he could figure it out, and he did.

Lenore, a newspaper columnist, wrote about her son's adventure, and two days later, she found herself in the midst of a media frenzy. Trial by fire; the verdict was clear. Headlines and interviews dubbed Lenore "The World's Worst Mom," a 2008 poster child for child neglect and endangerment in NYC. As I reflected on this story, I wondered about Lenore. Would she still defend her decision today? Would she be just as passionate about childhood independence—perhaps more so? Or did she shrink under the pressure, the judgment, the hatred?

I called her. (She hadn't shrunk.)

The Perils of a Non-Organic Grape

After her experience, Lenore recognized an urgent need to protect childhood independence and free play in our safety-obsessed culture. In the years that followed her story, she authored *Free-Range Kids* and cofounded the nonprofit Let Grow with renowned psychologists Peter Gray and Jonathan Haidt. "Somehow our culture has become obsessed with kids' fragility and has lost sight of their innate resilience," she tells me over the phone. "This concern grew out of good intentions! But treating kids as fragile is making them so. If you want to feed anxiety, just treat a competent person as incompetent. Warn them that everything's dangerous. Stop them from doing things they could handle."

Lenore speaks vibrantly, passionately about how a steady diet of fear-based media is shaping a culture of outlandish policies. In South Carolina, a mom asked the public elementary school to let her children—ages nine, ten, and eleven—walk the mile home on their own, and the school refused (Skenazy 2020b). In Brooklyn, a rabbi father let his kids (eleven and eight with a two-year-old in a stroller) walk a few blocks to the store and was arrested and charged with endangering the life of a child (Skenazy 2020a). In Louisiana, a fourth-grade boy was suspended for six days because the teacher glimpsed a BB gun in his room *at home* during a Zoom class (Monteverde 2020).

"We can't blame parents for helicoptering," Lenore says to me. "Our culture insists parents hover! But my goal is for it to be as normal to see a kid walking to the store as it is to see a kid in a car or on an iPad. Until we change our laws and norms, decent parents who want to nurture their children's growing capabilities will be forced to smother them instead." That's why Let Grow recently established a legal advocacy arm, passing the Reasonable Childhood Independence Law in eight states—and counting (Let Grow, n.d.). "Under the law, state child-welfare authorities can no longer take children away from their parents if they are demonstrating reasonably independent activities, as long as their kids are adequately fed, clothed, and cared for," Lenore explains.

Lenore cites "constant adult supervision, intervention, and

assistance" as the good intentions that often rob our kids of the experiences required to build resourcefulness and resilience. "What is anxiety if not the overwhelming belief that you cannot handle the world you live in? Kids need the chance to be on their own sometimes: playing, roaming, taking risks, getting scrapes, making things happen, and taking responsibility."

So what do we do? How do we free our households of an algorithm that feeds our fear? If technology and social media breed mistrust in our children, our families, and our society, how do we mend? Can we learn to opt out of ultrasurveillance and move toward a place of confidence, independence, trust?

"We need to reject the idea that kids are in constant physical, emotional, or psychological danger from creeps, kidnapping, frustration, failure, baby snatchers, bad grades, disappointing playdates, and/or the perils of a non-organic grape," Lenore says. To start, Lenore and her cofounders have created the Let Grow Project—a "homework" assignment teachers can give to students that says, "Go home and do something new, on your own, without your parents." The free, robust curriculum offers ongoing resources, project ideas, and goal sheets. An at-home version, the Let Grow Independence Kit, is free too.

+FOLLOW

You can find a link to access Let Grow's independence-building initiatives—and learn more about how to get involved in their compelling work—at *optoutfamily.com/more.*

Lenore also suggests schools start Let Grow Play Clubs, where school stays open for an hour or more before or after school for mixed-age, unstructured free play. No tech, no organized activities, no rules, except for one: don't hurt each other. Kids make their own fun. Hopscotch and kickball and tag. Piggyback rides. Cartwheels. Mud cakes. It's complementary time travel, shuttling today's anxious kids back to the playground experience of the 1980s.

But for lasting change to take place, we parents must lead the charge. We must examine our beliefs about how we guide our families. We must step into our roles as confident caregivers. We must trust ourselves, and we must trust our children.

"I realized the phone was giving me a false sense of security," one mother told me. "I was getting all of these notifications about what [my teen] was doing or where she was going or who she was texting. It was this barrier of trust, this thing that kept her from being autonomous. We both felt it. So we switched to flip phones. We're, I think, two months in, and it's harder than I thought. We've had to be more responsible. Writing down phone numbers, planning ahead, remembering our to-do lists, our deadlines. She had to buy an alarm clock. We keep an atlas in the car now. But we both feel free. She can finally grow up. She says it's been really empowering for her. I have to say exactly the same for me. It's like I went back in time. I feel young again. I guess you could say I'm an opt-out kid."

 DM

"Whenever I feel like I'm teetering on the edge of overparenting, I ask myself, 'What would an eighties parent do?' It helps me recalibrate knowing I was raised in a pretty loose time and I survived."
—Tina I.

I asked her daughter what the experience has been like for her. "I think the biggest thing that feels different is that now, when my mom asks me how my day was or, like, what I did, it feels like a real question. Like she actually cares and just wants to genuinely know, not like she's trying to catch me in a lie. Because before when she would ask me that question, I would be like, 'Mom, you already know because you tracked me all day.'"

A young adult I spoke with can relate all too well. "My mom still tracks me, like, all the time," she says. "A few days ago, she called me kind of frantic because I'd crossed state lines and didn't tell her. I had to remind her I'm twenty-four and moved out of the house five years ago."

Over Railroad Tracks, under Bridges

Many parents who track their children on apps like Life360—which is sold as a free "family safety app"—talk about how the world is too unsafe for them not to. But in fact, abductions by strangers are the rarest type of cases of missing children, with odds of their occurring similar to those of winning the lottery (Allen 2019). Still, we balk in disbelief at the thought of our kids roaming the neighborhood unsupervised. "Remember when we were kids?" said one mother I spoke with. "We'd be out riding bikes all over town—over railroad tracks, under bridges. Our kids can't do that anymore."

But in many ways, the physical world is safer for our kids than ever.

It's the mental world that requires immediate attention.

"There's stuff going on at school right now," says one teen I spoke with. "It's just, a lot. But I have to put on a brave face and pretend to be okay, because if I don't, my mom will take my phone away and things will get worse."

I asked other teens if they're sometimes nervous their parents will take away their phones. "Totally," said one. "If my mom knew about all the nude photos going around, she'd take my phone away for sure."

Another one said she delayed getting her parents involved when an older man was grooming her on a music platform. "I thought it was my fault," she said. "I wasn't supposed to have my phone in my room or even be online at that time, and I did it anyway. I didn't want to get in trouble, so I didn't tell them what was happening."

For these teens, the choice is simple: keep a secret or lose a phone?

■■■■■|||

So what happens next? How do we shield our kids from harmful content without shame, blame, or fallout? How do we course-correct, and when? And once we've encountered enough harmful content to pull the plug, how do we lessen the blow? Let's see what Meta's Trust and Safety team would do.

Transparency Center

"Since 2016, we've used a strategy called 'remove, reduce, inform' to manage content across Meta technologies," announces Meta in their updated Transparency Center (Meta, n.d.a). "This means we remove harmful content that goes against our policies, reduce the distribution of problematic content that doesn't violate our policies, and inform people with additional context so they can decide what to click, read, or share."

But Meta doesn't remove everything, all at once, in a grandiose gesture. Instead, the playbook calls for a tiered system that offers a spectrum of options for various scenarios, users, and circumstances. The good news? So, too, can we.

Remove

Want to give your child a totally tech-free experience? Ready to raise an opt-out kid? The simplest way to remove harmful content is not to provide access up front. By now, you'll know that giving any child a personal device introduces a slew of harms, stresses, and distractions from the vibrant life ahead of them. Plain and simple: remove the option. Don't purchase a smartphone for your child.

If you've already given your child a smartphone and you're unsure whether you've made the right decision, remember: you can course-correct at any time. You're in charge! The path ahead will be difficult, but not impossible. Removing your child's device is not a punishment or a consequence, it's simply an experiment. Wait for a moment when all is well and there's something lovely to look forward to. Perhaps your family is heading on vacation or you have concert tickets reserved or take-out is on the way. Explain to the family that you're all going to be experimenting with flip phones for a bit. Hand them out, one by one. Likely you'll be challenged. But your antifragility motto is this: this is not an emergency. Pop some popcorn, throw in a movie, and let everyone silently stew if they must. But wait a few days, a week. Help each other navigate finding a new rhythm as you run into barriers together. What parts feel hardest? Do you need to invest in alarm clocks? A point-and-shoot camera? A paper calendar? You're all in this together, as a team. Walk through it with intention—and each other.

Reduce

If you prefer a middle ground for your kids—an option some-where between tech free and tech dependent—the reduce approach offers plenty of flexibility. Maybe you're mostly comfortable with your family's relationship to technology, but every now and then you want to work out a few kinks and reset some unhealthy patterns. In this case, Meta has a few built-in strategies to try. "If content on Facebook doesn't violate the Facebook Community Standards, but might still be problematic or otherwise low-quality, Meta may reduce its distribution, consistent with user controls. This is one element of our broader 'remove, reduce, inform' strategy that we've used since 2016" (Meta 2023b).

So what does it look like to reduce the distribution of problem-atic content or tech use? Here's what Meta recommends:

Quiet Mode

"Quiet mode on Instagram turns off notifications and sends an auto-reply when someone sends you a direct message (DM), so you can focus on things like driving or studying" (Meta 2023a).

You can go dark. (Literally, if you'd like! Candlelight, pop-corn, couch forts, and spooky stories are a fam-wide favorite in our home.) Together, coordinate weekly times your family won't be using technology. Do you unplug every Sunday? Power down after 7:00 p.m.? Try a weekly sabbatical every few months? Whatever quiet mode you choose, do it together, and inform your close friends and family that you won't be accessible during those times. Remember, our patterns of technology use set an expectation for how available we will be at any moment. Be up front with boundaries and limitations. Then enjoy your freedom—together.

Block

Establish screen-free zones where your family won't be using personal devices: cars, dining rooms, bedrooms. One mother I know keeps it simple: "No technology while moving," she says. "Not only is it not safe, but it keeps long walks and car drives open for us to catch up with each other or, if we're alone, to contemplate and reflect on the day."

⏻ LOG OFF

Want to take this idea a step farther? Consider an opt-out getaway where all personal devices are left at home. (If preferred, designate a single phone to bring for emergency scenarios. Or don't! You choose.) Give yourself a week to settle into a more natural rhythm in the absence of pings, dings, blue lights, and badges. Put together a puzzle. Watch a sunset. Take a hike, a nap, a bubble bath. Stretch in the absence of power cords. And once you arrive home, see what permanent changes you'd like to invite into your everyday rhythm.

By creating spaces for connection and engagement, you're setting up habits and patterns that both reduce tech use and reclaim togetherness.

Report

If your children have access to a smartphone, it's inevitable that they will experience problematic content. In this case, take a cue from Meta and invite a formal report. Once your children have encountered anything alarming, harmful, or even tasteless, offer them the space to inform you. Meta simply offers one prompt: tell us more about what's wrong (Meta, n.d.b). You can give your child the same courtesy by asking questions, listening well, and thanking them for sharing with you. Once your child has finished processing, ask, "How can I help?" Together, form a plan.

Note: Meta also offers space to report anonymously. If you'd like, give your children the same sense of privacy by placing a box or a jar near a shared phone dock to collect anonymous struggles, questions, and curiosities.

Inform

If the remove and reduce methods feel too restrictive for your household's needs and values, you can (with utmost caution) choose a more relaxed approach. Perhaps you're working out a balance with a coparent or your children are preparing to leave the home to chart their own courses. As your household shapes its changing relationship to technology, Meta's inform stage can offer a handy guardrail against future issues. In this stage, Meta collaborates with "global experts in technology, public safety and human rights to create and update our policies" (Meta, n.d.a).

You can do the same. In addition to staying educated on the potential harms of technology—through conversations with friends, online research, and observations in your home—try these additional ideas to support your family's tech use:

Two-Factor Authentication

Two-factor authentication is an added safety measure when exchanging sensitive information online, such as passwords, account numbers, and email addresses. Channel this concept in your home by engaging the buddy system: If your child is uncomfortable talking through an online situation with you, suggest he or she take appropriate steps to enlist a backup. Maybe it's a trusted mentor, a counselor,

⊙ DM

"Before I allowed my teenage daughter to activate my old smartphone, I gave her two options: she could have a weekly tech check-in with her favorite aunt, or I would keep the phone. She chose her aunt."
—Whitney C.

a sibling, or a spiritual advisor. Decide together and put a plan in place before it's needed.

Quarterly Report

Each season, call for a pizza delivery and host a quarterly report to reassess the role technology is playing in your lives. Is everything working well? Does anyone have concerns? Who are you following? What are you learning? What's something that delighted you online? What needs to change? What doesn't feel good? Evaluate progress, check in with your family goals, and assess new plans you'd like to set in place. Then chow down!

By engaging your children in conversations and routines surrounding technology, you're acknowledging their voices and autonomy. But never relinquish the authority you hold as a parent. If something isn't right, take action. If you need to remove technology, remove it. But first, replace it with something even better.

It's Not AI, It's IA

"The definition of an IA [independent activity] is an unstructured, developmentally challenging task that is performed without any help from adults," writes Dr. Camilo Ortiz, who implemented Lenore's Let Grow framework into therapy for children (Ortiz 2023). "These activities are typically chosen by the child and fall into four categories (outdoor, indoor, with other children, involving mild risk of injury)." Riding a bike to the park, taking the subway alone. Cooking a meal, painting their bedroom walls. Going to a movie with friends—no supervision! Whittling, building a fire, camping in the back yard.

Dr. Ortiz developed IAs as a new treatment for kids who were experiencing high rates of anxiety. After noting that the best treatment plan he had wasn't working for the rising rates of child anxiety, he posited a new idea: intense and frequent childhood independence. "The beauty of this approach is that practically any-body can do it," he explains. "And you don't need to pay for some highfalutin PhD."

Dr. Ortiz notes that IAs are beneficial even when anxiety is specific to certain scenarios, such as a fear of dogs or sleeping alone. If a child is afraid of driving, for example, you need only say, "Okay, fine. What would you like to do? Windsurf? Make creme brulee? Spelunk? Run a marathon?" Let the child decide, and let the child go.

"IAs are 'topographically' different . . . from a child's anxieties. They are helpful because they exercise many of the psychological muscles needed to better tackle anxiety," Dr. Ortiz notes. "IAs increase resilience, tolerance of discomfort and uncertainty, social skills, and smart risk-taking."

His theory is astoundingly simple. And yet it works. Kids are combatting anxiety all over the country by reclaiming the very thing technology has stolen from them: childhood.

Tied the Logos On with Twine

It is spring.

Outside, the neighborhood is waking up from its winter slum-ber. The evidence is everywhere: chalked driveways, mulch bags, muddy boots. Inside, the tomato plants are ready—the ones my daughter started from seed in the depths of winter, lovingly cared for and sang to and raised well. There are peppers, too, and herbs. She has packaged fertilizer with care, tied the logos on with twine. "Add-ons are important," she says with a wink.

Today, she is asking whether she can pack up the little red wagon, whether she can walk around the neighborhood to sell her plants. She is asking whether she and her little brother can go alone.

I look at Ken, and he at me.

We decide, yes, yes, of course, and we point out all of the

houses they can go to if there is an emergency. We tell them to stay together. Ken sneaks his phone into the bottom of my daughter's backpack, tells me it's just a precaution. "If they're not back in a few hours," he reasons, "we can ping their location using Find My iPhone."

But I am thinking of IAs and of all I have learned, and something feels amiss about tracking them through the neighborhood they call home.

I tell him later of my conversation with so many parents, so many children, so many families throughout the writing of this book. I tell him how much I despise the lie that our kids can't be kept safe without surveillance. I tell him of Lenore's definition: "What is anxiety if not the fear that you cannot handle this world?" I tell him how much I want our kids to trust our very real and alive world, but mostly I want our kids to trust their very real and alive selves.

"I know what to do," he says.

A few days later, we wave goodbye to three kids under ten as they venture to the park, armed with walkie-talkies, a bag of mangoes, and a pocket knife.

"Have fun! Over!" Ken belts into his walkie-talkie.

"Roger that!" they shout.

And off they go.

✏️ **SELFIE**

What are the community standards in your home? Meta's exhaustive list includes categories like authenticity, safety, protection, and dignity. Together brainstorm a few community standards to ensure that the content you consume aligns with your family's goals. Intentionality? Collaboration? Creativity? If you're feeling stuck, consider reviewing your core values in chapter 13 to help you discern whether your technology practices are best supporting your family's path.

Closing Invitation

"Turn Your Head"

Life is deep and simple, and what our society
gives us is shallow and complicated.
—Fred Rogers

The old has passed away; behold, the new has come.
—2 Corinthians 5:17

Deep breath.

Here is where the work begins. But don't fret. All that we agree has worked against us—discovery and challenge and autoplay and data and vision and social proof and influence and wonder—can work for us. We can liberate ourselves and our kids from trudging through a world shaped by a cold and calculated algorithm.

Have you ever noticed that the videos that go viral—liked, loved, shared by millions—are often the simplest moments? It's the toddler making a funny face after trying his mom's vegan muffin. It's a baby finding her ear for the first time. It's a mom realizing those juice-box flaps are there for a reason. (Hidden handles—who knew?) It's the funny thing your kid will say to you in the car this morning, this very mundane morning, a moment that might have been missed completely.

Toddler lisps, cornfield mazes, watermelon smiles—these moments are all available to us, anytime, everywhere. We will wake to one today, another tomorrow. But we must notice them. We must dare to catch our own viral moments, not only to catch them but to see them, to observe them, to recognize their merit and mirth.

May we never again scroll through someone else's life without attending to our own.

Without Media to Divide

My assumption as I began writing this book was that I would find a way to untangle the digital power cords that I thought were strangling today's families. Without screens to distract us, without algorithms to manipulate us, without media to divide us, surely we might better engage with one another, as families, as a society, as a world? Finally we would find the time, the space, the common ground to connect with one another.

I still believe we will.

But there's more.

The simplest solutions are often the hardest to implement. It is easy to say, "Ditch your phone!" It is far more difficult to say, "Ditch your phone so you can become more intimate with the people around you, so you can remove the barrier between you and the ones you love, the barrier to your life together."

The truth is, it's hard to resist checking out when our responsibilities and emotions and lists loom large. It is difficult not to escape behind a screen. But also it's difficult not to escape behind something else. If not screens, we have dozens more ways to keep ourselves disconnected from the rest of our households. We have busyness, for one. Merlot. Me time. Self pity, selfishness. Golf. Envy. Secrets.

What I wanted to be true is that I would remove technology from my life and find myself lighter, freer, a little more wise. And all of that is true.

But what is also true is that in moments of fear or frustration or confusion, I still find myself disengaging. It doesn't take a glowing screen for me to go dark. In the absence of technology, I can just as easily bury myself in productivity or distract myself with something "worthy," such as cookies to bake or laundry to fold or math to check.

As it turns out, I can shut down faster than I ever could log on.

It is certainly easier to be present without a smartphone, without hunching—bowing?—over a screen in such a way that my heart is closed off from everyone else in the room. I have found my posture changing, ever so slightly. I am upright. I am open. But also, I am exposed.

There is an intimacy, a nakedness, that arrives with engaging fully in the world and with your family. These are the people who know you. This is the spouse you annoy and delight, the one you have fought with, the one you have cried over, the one you have turned toward and away from and back to. These are your children, your love and labor, the ones you pray for and laugh with and read to and yell at. They know nearly all of you. They know you pick at your cuticles when you're stressed. They know that dance you do when you have to pee. They know your favorite tea, your favorite pajamas, your favorite librarian and song and tree. They know that when you say, "It takes a family to raise a family," they'd better pick up their socks.

And that's the whole of it, isn't it? You are known, fully, all the way—your short fuse, your morning breath. The times you have lost your temper, lost your keys, lost your way. Your heart holds a history of hurts, and they're all jammed in there with the absolute joys and uncertainties and memories and the never-agains and the almosts and the if-onlys. And to engage with that? Well, the algorithm sounds easier, doesn't it?

But I am learning to try. I am learning to sit down, to listen. To apologize. To challenge my kids, to influence them, to trust them, to observe them with delight and joy and wonder. To let them play, to let them learn. To sit. To listen. To be.

As it turns out, I needn't be more engaging for my kids. I need only to be engaged *with* them.

It's like pausing the ticker-tape parade in your mind, letting the confetti fall, the floats go by, the drumbeats slow, and you are seeing, finally, glaringly, what is. You will fidget at first. And then you won't.

The intrusions still arrive, with or without technology. *Did I unplug the hairdryer? Did I print the boarding pass? Do we have time for pancakes? Am I enough? Can I handle this? Is everything going to be okay?* But I am learning to quietly consent to them. I am learning to treat them a bit like one would an interrupting toddler or, say, an ill-timed sprinkler system. Here it is again. Shall we go on anyway?

Your Mind and, Mostly, Your Heart

I am reminded of a conversation I had with a Silicon Valley developer years ago who predicted, accurately, that Big Tech isn't after only your attention. It's after your companionship. Your mind and, mostly, your heart. Intimacy. Togetherness, closeness, understanding. Warmth.

At the time, this idea seemed impossible, laughable even. How could a digital device ever replace companionship?

But just last week, I listened to Tristan Harris and Aza Raskin, cofounders of the Center for Humane Technology, speak about AI advancements, and they said this: "In the engagement economy, [there] was the race to the bottom of the brain stem. In second contact [of AI], it will be a race to intimacy." Whomever—or whatever—holds the primary intimate relationship in your life wins (Center for Humane Technology 2023).

And tonight, after hearing of a newly married couple who outsourced their wedding vows to ChatGPT, I receive an email introducing me to Replika, the "AI Companion Who Cares." I click to read more, but the tag line is all I can stomach: "Always here to listen and talk. Always on your side."

I do some digging to find that a San Francisco–based startup created Replika as "a digital twin to serve as a companion for the lonely, or even, one day, a version of ourselves that can carry out all the mundane tasks that we humans have to do, but never want to" (Murphy 2019).

But we know the truth, you and I. There are many, many tasks in our very human lives that we don't want to carry out. Waking up early to drive your son to his 5:30 a.m. swim practice. Making small talk in the bleachers. Picking up cilantro on your way home. But somewhere along the way, these mundane tasks stack up to a life. Your favorite song comes on in the grocery store and you can't help busting out your karaoke moves with the cashier, and your son laughs and rolls his eyes and you remember what he looked like at every age that has passed—his dimples at three, the tousled hair at six, the tiny chip in his front tooth you never fixed because everyone grew to love it.

You forget the cilantro, but my gosh, the sunrise looks so beautiful in the parking lot.

Turn Your Head

Early this week, in the front yard, my toddler senses she does not have my full attention. We are pulling weeds, but I am thinking of deadlines and a conversation that went awry the day prior, and I'm mentally reminding myself to thaw beef for tacos.

"Turn your head," she tells me.

She grabs my head gently, her dirt-caked hands on both of my cheeks. She has learned to do this when I'm not hearing her—or perhaps better put, when I'm not listening. It's what she does when there's something she'd like me to see that she can't fully verbalize or that's far too important to miss.

"Turn your head."

I do, and I catch a glimpse of two children bounding onto our rope swing, their laughter caught in the wind. Their hair blows wildly and the sun is high, and for a moment, my daughter and I listen without words. As I watch them sway and chatter and dream, I am filled with hope for the generation to come that will once again be given permission to delight in a cloudless sky, a tender playmate, an afternoon of joy to swing upside down among the leaves. For a growing movement of children who will dance without TikTok. Who will live and love without commentary. Who will smile— widely and freely—without a filter.

And I am filled with gratitude for the bold, brave, and unapologetic parents who will point their children to the heights of presence and admiration and innocence that Silicon Valley can never scale.

As I gather up the last of the weeds we've pulled, as my daughter and I drink our water cold from the garden hose, she spots a small yellow caterpillar inching over a tomato leaf. And it occurs to me that I may never learn to be more engaging than the algorithm. I may never succeed in rewriting the devious playbook that guides our society to join a faceless army of complacency, distraction, burnout.

But it won't matter. The heart of a child arrives with a playbook all of its own—a better one, one that's far more captivating than a hoodied tech bro could ever conjure. We need only pay attention, to listen hard, to stay close as the lessons unfold with time.

Becoming an opt-out family isn't all about clenching our fists and teeth and plans as we strategize 101 ways to conquer the gods of tech. It's not about wringing our hands in worry, furrowing our brows in anger, rolling our eyes in judgment. It's not about fearfully shooing our children away from screens, slamming laptops shut in fits of despair.

It is, simply, opening the door wide to something better.

It's the quiet recognition that every time we opt out of technology, we opt in to life. It's holding out our palms, bending low, bearing witness to a crawling caterpillar and the lispy toddler who has just named him Howard.

It's pausing to notice the things technology begs us not to: the soil beneath us and the cardinal above us, but mostly the beating hearts before us. It's offering a thank you, and a prayer, as we lift our eyes to see the bigger picture that awaits on the other side of the screen.

After all, the opposite of opting in is not opting out.

It's living free.

The Opt-Out Family Pledge

In our household, we reject the lie that technology is neutral. We celebrate the present moment we are given. We pause to listen attentively to the hearts of others. We seek knowledge with purpose and patience, knowing that the first answer is not always the right one. We face life's challenges. We make our own fun. We rest in the truth that easier is not better and that different is not wrong. We observe context in all things. We remember that influence is earned. We build habits that support our values.

In our household, we will live in the understanding that people are more important than phones.

We will look up. We will chase mystery. We will be free.

Acknowledgments

First things first: to you, to you, to you. To every family who is unafraid to swim upstream, to try a new way, to rejigger the status quo. Wherever you land on this great wide tech spectrum (techtrum?), it is a great honor to join you in questioning the virtues of things we've been told to behold.

There are many families I have learned a few things from, and a few families I have learned much from. To my own parents, who uttered the famed 1980s rejoinder, "If all of your friends jumped off a bridge, would you?" enough times to stick. To the Staytons, who taught me that being weird is usually a sign that you're on the right track. To the Hansens, who opened the door to a dream we didn't know we had. To the Rothenbushes, who believe deeply in dessert. To the Mahurons, who taught me that speaking in a British flight-attendant accent keeps children from getting carsick on bumpy rides. To the Woolners, who showed me that faith and authenticity needn't ever be at odds. To the Kidstons, who never miss a birthday—or a chance to celebrate. To the Thomas family, who always keeps a spare pack of Monopoly Deal at the dinner table. To the Bredows, who taught me about cookies—both in code and chocolate chip.

To our Rock family: you are missed. To our Sonrise family:

you are loved. To the Watsons, the Hills, and the Orlikowskis: be careful, I have Arethra.

To my two older sisters, who taught me the most engaging trick I know: when in doubt, play water fountain.

To my own pint-sized Luddite Club: Mallory, Charlie, Olive, Roman, Rachel, Cora, Joanna, Hannah, Jesse, Erin, Alexa, Caleb, Elsie: may the trampoline someday hold us all.

To Caleb Stayton for good advice that changed everything.

To Heather Sommer for teaching me joy, laughter, and spatchcocking. There's no one I'd rather risk capsizing a kayak with than you.

To Sheri Carlstrom, one of few people who can get me to pick up the phone on a Saturday afternoon: you have turned my kitchen counter into a church pew, many times over. I am so grateful for your wisdom.

To Ainsley Arment, who created a community of motley misfits that—somehow—all belong to each other: thank you for welcoming me into your fold.

To Carla Beard for teaching me how to write. To Laura O'Hara for teaching me how to fall in love with writing. To Margie Yeager for teaching me to delete everything else.

To the Tills for caffeine and cheer. To the Pratts for saying yes to the basement. To the Coateses for butterfly-catching and leaf sticks. To the Floors for rainy park mornings and hot tea. To the St. Johns for many years spent rattling the barn. To Jennifer Wright for chair pose. To Elissa Watts, one of few reasons I miss Voxer.

To Shu-Hui, who gave our kids what tech never could—in another language, at that. To Terry and Lisa Ellis, Sally McGovern, Sara Scantlin, Holly Hemsoth, Kris Botas, and everyone at Fire and Light who guides our kids to imagine greater stories. To ATA Coventry for bringing "ma'am" back.

To the Parsleys, Truesdales, Forces, and Rushings for letting our kids run freely through our neighborhood. You are wonderful neighbors, even better friends.

To Ginny Yurich, Ashlee Gadd, Jessica Elefante, Tsh Oxenreider, Ellie Holcomb, Jenna Kutcher, Sarah Mackenzie, Sally Clarkson, Mallory Ervin, Julie Bogart, Shanna Skidmore, Denaye Barahona, and many more who champion my work even after I have no retweets or Insta stories to offer you in return: thank you for believing in the me beyond the platform.

To Maria Shriver for Los Angeles and New York.

To Gabrielle Blair and the early ALT team for offering me my first mic of many. To Scripps Network and the HGTV.com crew: thanks for giving me a reason to keep wiping the drywall dust off my laptop.

To Steven Kurutz and the *New York Times* for the phone call I almost missed.

To the dear friends who invited me to learn the life-altering lessons a new culture provides. To Barrett, Marisa, and ABLE: thank you for Ethiopia. To Carly: thank you for Haiti. To Jyoti, Raj, and Priti: thank you for India. To Danny and Mara: thank you for Ecuador. To Rob and Jill: thank you for London. To Katie: thank you for Ireland. To Irene: thank you for Singapore.

To the digital sages who delicately unloose tech's tangled web: Geoffrey Fowler, Jaron Lanier, Tristan Harris, Aza Raskin, Frances Haugen, Kevin Kelly, Adam Alter, Susan Linn, Nicholas Kardaras, Jean Twenge, Sherry Turkle, Jonathan Haidt, Josh Golin.

To the sources of wisdom who offered me endless hours of research, guidance, and care: Pam Leo, Erin Walsh, Dr. Carla Hannaford, Chris McKenna, Kathryn Starke, Dr. Victoria Dunckley, Titania Jordan and the team at Bark Phone, Joann Bogard, Sharon Winkler, Catherine Steiner-Adair, Jason Frost, Joe Clement, Matt Miles, Jean Rogers and the Screen Time Action Network at Fairplay, Mathew Georghiou, Drea Burbank, Dr. Larry Cohen, Travis Bauman, Dr. Jean Lomino, Dr. Wendy Shofer, Brooke Shannon and Wait Until 8th movement, Joey Odom and the Aro team, Lenore Skenazy and Let Grow, Kim John Payne, and Dr. Kevin Gary. A million thanks. You have enriched this book in every sense of the word.

To the fearless team at Zondervan who didn't bat an eye when I said I'd like to try publishing this book the old school way. To Carly Kellerman for giving it the green light. To Carolyn McCready for driving it home. To Margot Starbuck for everything in the middle—and for forgiving these chapters for the infamous phone-free bike crash.

To the faithful readers at Design for Mankind: thank you for sticking around for twenty years! Your presence keeps me astonished; your comments keep me wise. To the plucky community at Other Goose: it's an honor to homeschool alongside you.

To Betsy Loechner for countless gifts over many years: your ribs recipe, construction aid, bench cushions turned book forts, Polynesian pool detours. If our children grow up thinking you've hung the moon, they're not too far off the truth.

To my husband, Ken: my love and respect for you are impossible to sum up in a few lines or less. (And I know you won't read this anyway!) So I will simply add this for anyone else who is paying attention: had I known how well you would guide our family, how bravely and boldly you'd fight with and for us, how deep your integrity, how strong your convictions, how quick your wit, and how capable your leadership, I'd have run to the altar even faster. Here's to our third act, smidge.

To my children, who inspire my best ideas and interrupt my worst ones: this one's for you.

References

AACAP. 2020. "Screen Time and Children." www.aacap.org/AACAP /Families_and_Youth/Facts_for _Families/FFF-Guide/Children-And -Watching-TV-054.aspx.

Akhtar, Allana, and Marguerite Ward. 2020. "Bill Gates and Steve Jobs Raised Their Kids with Limited Tech—and It Should Have Been a Red Flag about Our Own Smartphone Use." *Business Insider* (May 15). www.businessinsider .com/screen-time-limits-bill-gates -steve-jobs-red-flag-2017-10#even-elite -silicon-valley-schools-are-noticeably -low-tech-5.

Allen, Felix. 2020. "Dying for Likes: Dark Truth of Social Media as US Pre-Teen Girl Suicides Soar 150% and Self-Harm Triples, Netflix's Social Dilemma Reveals." *US Sun* (September 17). www.the-sun.com /news/1487147/social-media-suicides -self-harm-netflix-social-dilemma/.

Allen, Jonathan. 2019. "Kidnapped Children Make Headlines, but Abduction Is Rare in U.S." Reuters (January 11). www.reuters.com /article/us-wisconsin-missinggirl -data/kidnapped-children-make-head lines-but-abduction-is-rare-in-u-s -idUSKCN1P52BJ.

Alter, Adam. 2018. *Irresistible: The Rise of Addictive Technology and the Business of Keeping Us Hooked.* New York: Penguin.

Amazon. n.d. "Alexa Features." Amazon.com. www.amazon.com/b?ie =UTF8&node=21576558011.

American Psychological Association. 2015. "Frequently Monitoring Progress toward Goals Increases Chance of Success." American Psychological Association. www.apa.org/news/press /releases/2015/10/progress-goals.

Arundel, Kara. 2023. "'Wave' of Litigation Expected as Schools Fight Social Media Companies." K-12 Dive (June 1). www.k12dive.com/news /wave-of-social-media-litigation -expected/651224/.

Auxier, Brooke, Monica Anderson, Andrew Perrin, and Erica Turner. 2020a. "4. Parents' Attitudes—and Experiences—Related to Digital Technology." Pew Research Center. www.pewresearch.org/internet/2020/07/28/parents-attitudes-and-experiences-related-to-digital-technology/.

Auxier, Brooke, Monica Anderson, Andrew Perrin, and Erica Turner. 2020b. "Parenting Kids in the Age of Screens, Social Media and Digital Devices." Pew Research Center. www.pewresearch.org/internet/2020/07/28/parenting-children-in-the-age-of-screens/#fn-26219-3.

Bark. 2023. "Spotify Has a Porn Problem—Here's What Parents Need to Know." Bark Blog (January 27). www.bark.us/blog/spotify-porn-problem/.

BBC. 2020. "TikTok Holocaust Trend 'Hurtful and Offensive.'" BBC (August 27). www.bbc.com/news/newsbeat-53934500.

BBC. 2022. "Molly Russell Inquest: Pinterest Executive Admits Site Was Not Safe." BBC (September 22). www.bbc.com/news/uk-england-london-62991510.

Belkin, Lisa. 2009. "What Is Slow-Parenting?" *New York Times* (April 8). https://archive.nytimes.com/parenting.blogs.nytimes.com/2009/04/08/what-is-slow-parenting/.

Bernhardt-Lanier, Celine, and Aliza Kopans. n.d. "Dear Parents: A Digital Well-Being Resource from Teens to Parents." Children's Screen Time Action Network. https://screentimenetwork.org/sites/default/files/resources/Dear%20Parents.pdf.

Bernstein, Gaia. 2023. "Unwired: Gaining Control over Addictive Technologies." Next Big Idea Club (April 14). https://cdn3.nextbigideaclub.com/magazine/unwired-gaining-control-addictive-technologies-bookbite/41632/bb_-gaia-bernstein_mix/amp/.

Biden, Joe. 2023. "President Biden Remarks on Banking and the Economy." C-SPAN (March 13). www.c-span.org/video/?526672-1/president-biden-americans-confidence-us-banking-system.

Big Brains. 2022. "Terms and Conditions." Big Brains. Accessed December 14, 2023. https://brainapplabs.com/terms-of-use/.

Brandom, Russell. 2019. "AI Pioneer Accused of Having Sex with Trafficking Victim on Jeffrey Epstein's Island." *The Verge.* www.theverge.com/2019/8/9/20798900/marvin-minsky-jeffrey-epstein-sex-trafficking-island-court-records-unsealed.

Britannica. 2023. s.v. "Algorithm." www.britannica.com/science/algorithm.

Brown, Abram. 2022. "Saint Frances of the Whistle: A Year after Frances Haugen's Facebook Leak, Every Corporate Secret Is Now Up for Grabs." The Information. www.theinformation.com/articles/saint-frances-of-the-whistle-a-year-after-frances-haugens-facebook-leak-every-corporate-secret-is-now-up-for-grabs.

Buolamwini, Joy. 2016. "InCoding—In the Beginning Was the Coded Gaze." *Medium* (May 16). https://medium .com/mit-media-lab/incoding-in-the -beginning-4e2a5c51a45d.

Burnham, David. 1984. "The Computer, the Consumer and Privacy." *New York Times* (March 4). www.nytimes.com/1984/03/04 /weekinreview/the-computer-the -consumer-and-privacy.html.

Buscho, Ann Gold. 2021. "Marriage, Divorce, and Social Media: A Recipe for Disaster." *Psychology Today* (November 9). www.psychologytoday .com/us/blog/better-divorce/202111 /marriage-divorce-and-social-media -recipe-disaster.

Carman, Tessa. 2021. "Children Are Born Persons." Great Hearts Institute. https://greathearts.institute/children -are-born-persons/.

CDC. 2008. "CDC Study Warns of Deaths Due to the Choking Game." CDC (February 14). www.cdc.gov /media/pressrel/2008/r080214.htm.

Center for Humane Technology. 2023. "The A.I. Dilemma—March 9, 2023." YouTube (April 5). www.youtube.com /watch?v=xoVJKj8lcNQ.

Chen, Angela. 2020. "Eight Revelations from MIT's Jeffrey Epstein Report." *MIT Technology Review* (January 10). www.technologyreview .com/2020/01/10/130928/mit-jeffrey -epstein-donations-media-lab-joi-ito -seth-lloyd-funding-ethics.

Cherry, Amy. 2020. "New Form of Cyberbullying Taking Hold in the Virtual Learning Environment." WDEL (November 2). www.wdel.com /news/new-form-of-cyberbullying -taking-hold-in-the-virtual-learning -environment/article_5db520fa-1d36 -11eb-b65a-c3440d637dab.html.

Chiara. 2023. "TikTok's 'Focused View': The Creepy New Feature Aims to Monetize Your Emotions." Access Now (February 7). www.accessnow .org/tiktoks-focused-view-creepy -feature-monetise-your-emotions-2/.

Chicago Review Press. 2017. "*Screen Schooled* Authors Joe Clement and Matt Miles Discuss Technology in the Classroom." CRP's Blog (August 18). www.chicagoreviewpress.com/blog /screen-schooled-authors-joe-clement -and-matt-miles-discuss-technology-in -the-classroom-whats-really-going-on-- think-distracted-kids-with-poor-problem -solving-skills-and-little-intelle/.

Children's Safety Network. n.d. "Drowning Prevention." Children's Safety Network. Accessed August 19, 2023. www.childrenssafetynetwork .org/child-safety-topics-terms /drowning-prevention.

Choudary, Sangeet Paul, 2014. "Reverse Network Effects: Why Today's Social Networks Can Fail as They Grow Larger." *Wired* (March). www.wired.com/insights/2014/03 /reverse-network-effects-todays-social -networks-can-fail-grow-larger/.

Clark, Mitchell. 2021. "Alexa Told a Child to Do Potentially Lethal 'Challenge.'" *The Verge* (December 28). www.theverge.com/2021/12/28 /22856832/amazon-alexa-challenge -child-dangerous-electricity-algorithm.

Clarke, Katrina. 2014. "Study Finds Thousands of Teens Going to Extreme Lengths for Online Attention." *Toronto Star* (November 12). www.thestar.com/news/gta/2014/11/12/study_finds_thousands_of_teens_going_to_extreme_lengths_for_online_attention.html.

Clifford, Catherine. 2018. "You Can Pay for Your Burger with Your Face at This Fast Food Restaurant, Thanks to A.I." CNBC (February 2). www.cnbc.com/2018/02/02/pay-with-facial-recognition-a-i-at-caliburger-in-pasadena-california.html.

Cobb, Jelani. 2022. "Why I Quit Elon Musk's Twitter." *New Yorker* (November 27). www.newyorker.com/news/daily-comment/why-i-quit-elon-musks-twitter.

Cohen, Lawrence J. 2023. "Finding the Line between Supporting and Rescuing Kids." *Psychology Today* (February 19). www.psychologytoday.com/intl/blog/playful-parenting/202302/finding-the-line-between-supporting-and-rescuing-kids.

Collins, James C. 2001. *Good to Great: Why Some Companies Make the Leap . . . and Others Don't.* New York: HarperBusiness.

Comaford, Christine. 2020. "Are You Getting Enough Hugs?" *Forbes* (August 22). www.forbes.com/sites/christinecomaford/2020/08/22/are-you-getting-enough-hugs/.

Common Sense Media. 2023. "New Report Reveals Truths about How Teens Engage with Pornography." Common Sense (January 10). www.commonsensemedia.org/press-releases/new-report-reveals-truths-about-how-teens-engage-with-pornography.

Cost, Ben, Asia Grace, Marisa Dellatto, Eric Hegedus, and Sophie Gardiner. 2023. "Twenty-Five Craziest TikTok Challenges and the Ordeals They've Caused." *New York Post* (July 27). https://nypost.com/article/craziest-tiktok-challenges-so-far/.

Coughlan, Sean. 2018. "'Sharenting' Puts Young at Risk of Online Fraud." BBC (May 20). www.bbc.com/news/education-44153754.

Cowan, Brian. n.d. "Joseph Addison." *Digital Encyclopedia of British Sociability in the Long Eighteenth Century.* Accessed October 9, 2023. www.digitens.org/en/notices/joseph-addison.html.

Csikszentmihalyi, Mihaly. 2008. *Flow: The Psychology of Optimal Experience.* New York: HarperPerennial.

Davids, Surur. 2018. "Microsoft and National Australia Bank Brings Windows Hello to ATMs." MSPowerUser. https://mspoweruser.com/microsoft-and-national-australia-bank-brings-windows-hello-to-atms/.

Davies, Dave. 2022. "Users Beware: Apps Are Using a Loophole in Privacy Law to Track Kids' Phones." NPR. www.npr.org/2022/06/16/1105212701/users-beware-apps-are-using-a-loophole-in-privacy-law-to-track-kids-phones.

De Witte, Melissa. 2022. "Gen Z Are Not 'Coddled.' They Are Highly Collaborative, Self-Reliant and Pragmatic, according to New Stanford-Affiliated Research." Stanford News.

news.stanford.edu/2022/01/03/know
-gen-z/.

Dickson, EJ. 2020. "TikTok Stars Are
Being Turned into Deepfake Porn
without Their Consent." *Rolling Stone*
(October 26). www.rollingstone.com
/culture/culture-features/tiktok
-creators-deepfake-pornography
-discord-pornhub-1078859/.

Dimson, Thomas. 2022. "Machine
Learning at Scale." Facebook. www
.facebook.com/atscaleevents/videos
/1856120757994353/.

Donnelly, Laura. 2019. "Children
Spend Twice as Long on Smartphones
as Talking to Parents." *Telegraph*
(February 7). www.telegraph.co.uk
/news/2019/02/07/children-spend
-twice-long-smartphones-talking
-parents/.

Dru, Jean-Marie. 1996. *Disruption:
Overturning Conventions and Shaking Up
the Marketplace*. Translated by Robin
Lemberg, Sarah Baldwin, and Jean-
Marie Dru. New York: Wiley.

Duarte, Fabio. 2023. "Number of
ChatGPT Users (2023)." Exploding
Topics. https://explodingtopics.com
/blog/chatgpt-users.

Dunckley, Victoria L. 2016. "Restricting
Screens: Why Your Child WON'T
Get Left Behind." *Psychology Today*
(March 22). www.psychologytoday
.com/intl/blog/mental-wealth/201603
/restricting-screens-why-your-child
-wont-get-left-behind.

Economist. 2022. "Covid Learning
Loss Has Been a Global Disaster."
Economist (July 7). www.economist

.com/international/2022/07/07/covid
-learning-loss-has-been-a-global-disaster.

Ellerbeck, Stefan. 2022. "Half of
US Teens Use the Internet 'Almost
Constantly.' But Where Are They
Spending Their Time Online?" World
Economic Forum (August 30). www
.weforum.org/agenda/2022/08/social
-media-internet-online-teenagers
-screens-us/.

Faber, Adele, and Elaine Mazlish.
2006. *How to Talk So Teens Will Listen
and Listen So Teens Will Talk*. New York:
HarperCollins.

Fairplay. 2021. "Feb. 19, 2021.
Advocates to FTC: Prodigy Math
Game Preys on Kids and Families."
Fairplay. https://fairplayforkids.org
/feb-19-2021-advocates-to-ftc-prodigy
-math-game-preys-on-kids and--
families/.

Fairplay. n.d. "Your Kids and
Instagram Youth: What Your Family
Needs to Know." Fairplay. Accessed
August 18, 2023. https://fairplayforkids
.org/pf/instagram-and-your-kids/.

Fisher, Miles. 2022. "How I
Became the Deepfake Tom Cruise."
Hollywood Reporter (July 21). www
.hollywoodreporter.com/business
/digital/deepfake-tom-cruise-miles
-fisher-1235182932/.

Florio, Erin. 2018. "At Marriott, You
Can Now Check In with Your Face."
Condé Nast Traveler (July 11). www
.cntraveler.com/story/marriott-alibaba
-facial-recognition-hotel-check-in.

F0t0b0y. 2017. "Amazon Alexa Gone
Wild!!! Full Version from Beginning

to End." YouTube. www.youtube.com /watch?v=epyWW2e43UU.

Fowler, Geoffrey A. 2022. "Algorithms Prey on You. What If You Could Reset Them?" *Washington Post* (May 12). www.washingtonpost.com /technology/2022/05/12/instagram -algorithm/.

Fry, Hannah. 2019. *Hello World: Being Human in the Age of Algorithms.* New York: Norton.

Gabb. 2022. "Technology Addiction in Kids." Gabb.com (January 25). https:// gabb.com/blog/kids-addicted-to -technology/.

GAO. 2022. "Facial Recognition Technology: CBP Traveler Identity Verification and Efforts to Address Privacy Issues." US Government Accountability Office (July 27). www .gao.gov/products/gao-22-106154.

Garcia-Navarro, Lulu. 2023. "The Teenager Leading the Smartphone Liberation Movement." *New York Times* (February 2). www.nytimes.com /2023/02/02/opinion/teen-luddite -smartphones.html?.

Gartner. n.d. "Gartner Hype Cycle Research Methodology." Gartner. Accessed August 18, 2023. www.gartner.com/en/research /methodologies/gartner-hype-cycle.

Ghaffary, Shirin, and Alex Kantrowitz. 2021. "'Don't Be Evil' Isn't a Normal Company Value. But Google Isn't a Normal Company." *Vox* (February 16). www.vox.com/recode /2021/2/16/22280502/google-dont-be -evil-land-of-the-giants-podcast.

Google. n.d. "About Presence Sensing and How to Manage Your Data." Google Nest Help. Accessed August 18, 2023. https://support .google.com/googlenest/answer /10000312?hl=en.

Gopnik, Alison, Andrew N. Meltzoff, and Patricia K. Kuhl. 2001. *The Scientist in the Crib: What Early Learning Tells Us about the Mind.* New York: HarperCollins.

Gordon, Allison, and Pamela Brown. 2023. "Surgeon General Says Thirteen Is 'Too Early' to Join Social Media." CNN (January 29). www.cnn.com /2023/01/29/health/surgeon-general -social-media/index.html.

Graham, Kristen A. 2022. "Philly Schools Will Vote to Spend $5 to Keep Students' Cell Phone Locked Up." *Philadelphia Inquirer* (October 19). www.inquirer.com/news/philadelphia -schools-cell-phone-yondr-pouch -20221019.html.

Gray, Peter. 2015. *Free to Learn: Why Unleashing the Instinct to Play Will Make Our Children Happier, More Self-Reliant, and Better Students for Life.* New York: Basic Books.

Grothaus, Michael. 2023. "Facebook and Instagram Paid Verification Will Allow Anyone to Get a Blue Check." Fast Company (February 2). www .fastcompany.com/90853370/meta -paid-verification-facebook-instagram -zuckerberg-subscription.

Hackett, Sam. 2023. "The Complete Timeline of Instagram Updates That Have Changed the Way We Gram." Kicksta blog. https://blog.kicksta.co

/the-complete-timeline-of-instagram
-updates/.

**Hadar, Aviad, Itay Hadas, Avi
Lazarovits, Uri Alyagon, Daniel
Eliraz, and Abraham Zangen.**
2017. "Answering the Missed Call:
Initial Exploration of Cognitive
and Electrophysiological Changes
Associated with Smartphone Use
and Abuse." PLOS. (July 5). https://
journals.plos.org/plosone/article?id=10
.1371/journal.pone.0180094.

Haggarty, Elizabeth. 2010. "Speed
Camera Lottery Pays Drivers
for Slowing Down." *Toronto Star*
(December 9). www.thestar.com/news
/world/2010/12/09/speed_camera
_lottery_pays_drivers_for_slowing
_down.html.

Hannaford, Carla. 2005. *Smart Moves:
Why Learning Is Not All in Your Head.*
Second edition. Salt Lake City: Great
River Books.

Hanson, Alec. 2023. "The Future
of Mortgage Lending: How AI
And Humans Can Coexist." *Forbes*
(March 9). www.forbes.com
/sites/forbesfinancecouncil/2023
/03/09/the-future-of-mortgage
-lending-how-ai-and-humans-can
-coexist/?sh=43b63554a625.

Hanssen, Bjørn-Rune. 2017. "The
Four Freedoms of Games and
Gamification." Puzzel (January 20).
www.puzzel.com/2017/01/20/four
-freedoms-games-gamification/.

Hari, Johann. 2022. "Your Attention
Didn't Collapse. It Was Stolen."
Guardian (January 2). www
.theguardian.com/science/2022/jan

/02/attention-span-focus-screens-apps
-smartphones-social-media.

Harris, Tristan. 2023. X. https://
twitter.com/tristanharris/status
/1634299911872348160?s=46&t
=dKA6bnISNJikb2vECGEsKw.

Harwell, Drew. 2022. "Remote
Learning Apps Shared Children's Data
at a 'Dizzying Scale.'" *Washington
Post* (May 24), www.washingtonpost
.com/technology/2022/05/24/remote
-school-app-tracking-privacy/.

Haught, Nancy. 2012. "Barry Glassner
Calls on Americans to Face Down
'the Culture of Fear." *Oregonian*
(January 25). www.oregonlive.com
/living/2012/01/barry_glassner_calls
_on_americ.html.

Haydon, Kathryn P. 2018. "Should We
All Take Note When Silicon Valley
Parents Opt out of Schools with Tech?"
Sparkitivity (January 3). https://
sparkitivity.com/2018/01/03/should
-we-all-take-note-when-silicon-valley
-parents-opt-out-of-schools-with-tech/.

Hayes, Adam. 2023. "YouTube Stats:
Everything You Need to Know in
2023!" Wyzowl (September 20). www
.wyzowl.com/youtube-stats/.

Helmore, Edward. 2023. "Why Did
the $212bn Tech-Lender Silicon Valley
Bank Abruptly Collapse?" *Guardian*
(March 17). www.theguardian.com
/business/2023/mar/17/why-silicon
-valley-bank-collapsed-svb-fail.

Hernandez, Joe. 2022. "A Parents'
Lawsuit Accuses Amazon of Selling
Suicide Kits to Teenagers." NPR
(October 9). www.npr.org/2022/10/09

/1127686507/amazon-suicide-teenagers
-poison.

Herold, Benjamin. 2016. "Facebook's
Zuckerberg to Bet Big on Personalized
Learning." *Education Week* (March 7).
www.edweek.org/policy-politics
/facebooks-zuckerberg-to-bet-big-on
-personalized-learning/2016/03.

Herold, Benjamin. 2019. "Forty
Percent of Elementary School Teachers'
Work Could Be Automated by 2030,
McKinsey Global Institute Predicts."
Education Week (June 4). www.edweek
.org/education/forty-percent-of
-elementary-school-teachers-work
-could-be-automated-by-2030-mckinsey
-global-institute-predicts/2019/06#.

Hollister, Sean. 2020. "Your Philips
Hue Light Bulbs Can Still Be Hacked—
and Until Recently, Compromise Your
Network." *The Verge* (February 5).
www.theverge.com/2020/2/5
/21123491/philips-hue-bulb-hack-hub
-firmware-patch-update.

Howells, Kristina. n.d. "Sweden
Rewards Good Drivers." *Akashic Times.*
Accessed August 18, 2023. https://
akashictimes.co.uk/sweden-rewards
-good-drivers/.

Huddleston Jr., Tom. 2022. "TikTok
Shares Your Data More Than Any Other
Social Media App—and It's Unclear
Where It Goes, Study Says." CNBC
(February 8). www.cnbc.com/2022/02
/08/tiktok-shares-your-data-more-than
-any-other-social-media-app-study.html.

**Hutton, John S., Jonathan Dudley,
and Tzipi Horowitz-Kraus.** 2019.
"Associations between Screen-Based
Media Use and Brain White Matter

Integrity in Preschool-Aged Children."
JAMA Pediatrics (November 4). doi.org
/10.1001/jamapediatrics.2019.3869.

Iovine, Anna. 2023. "What Are
Parasocial Relationships?" Mashable
(September 7). https://mashable.com
/article/parasocial-relationships
-definition-meaning.

Iqbal, Mansoor. 2023. "Home App
Data Instagram Revenue and Usage
Statistics (2023)." Business of Apps.
www.businessofapps.com/data
/instagram-statistics/.

Jiang, Jingjing. 2018. "How Teens and
Parents Navigate Screen Time and
Device Distractions." Pew Research
Center. www.pewresearch.org
/internet/2018/08/22/how-teens-and
-parents-navigate-screen-time-and
-device-distractions/.

Kadane, Lisa. 2022. "Positive
Reinforcement Is the One Parenting
Trick Everybody Needs to Know."
Today's Parent (May 5). www
.todaysparent.com/family/parenting
/positive-reinforcement-one-parenting
-trick-everybody-needs/.

Kale, Sirin. 2018. "Logged Off: Meet
the Teens Who Refuse to Use Social
Media." *Guardian* (August 29). www
.theguardian.com/society/2018/aug
/29/teens-desert-social-media.

Kane, Thomas. 2022. "Kids Are
Far, Far Behind in School." *Atlantic*
(May 22). www.theatlantic.com/ideas
/archive/2022/05/schools-learning
-loss-remote-covid-education/629938/.

Kardaras, Nicholas. 2016. "It's 'Digital
Heroin': How Screens Turn Kids into

Psychotic Junkies." New York Post (August 27). https://nypost.com/2016 /08/27/its-digital-heroin-how-screens -turn-kids-into-psychotic-junkies/.

Kardaras, Nicholas. 2017. *Glow Kids: How Screen Addiction Is Hijacking Our Kids—and How to Break the Trance.* New York: St. Martin's.

Kelly, Kevin. 2016. *The Inevitable: Understanding the Twelve Technological Forces That Will Shape Our Future.* New York: Viking.

Knox, Liam. 2023. "Admissions Offices Deploy AI." Inside Higher Ed (October 9). www.insidehighered.com /news/admissions/traditional-age /2023/10/09/admissions-offices-turn -ai-application-reviews.

Kuo, Lily. 2019. "In China, They're Closing Churches, Jailing Pastors—and Even Rewriting Scripture." *Guardian* (January 13). www.theguardian.com /world/2019/jan/13/china-christians -religious-persecution-translation-bible.

Lanier, Jaron. 2019. *Ten Arguments for Deleting Your Social Media Accounts Right Now.* New York: Picador.

Lea, Jessica. 2019. "Girls in the Same Youth Group Targeted by Sex Offender via Bible App." ChurchLeaders (November 5). https://churchleaders .com/news/365325-girls-youth-group -targeted-registered-sex-offender-bible -app.html.

Learning Network. 2020. "What Students Are Saying about How Much They Use Their Phones, and Whether We Should Be Worried." *New York Times* (February 6). www.nytimes.com /2020/02/06/learning/what-students -are-saying-about-how-much-they-use -their-phones-and-whether-we-should -be-worried.html.

Lee, Catherine. 2022. "Charlotte Mason: Education Pioneer Was 'Guiding Light.'" BBC News (January 1). www.bbc.com/news/uk -england-cumbria-59124124.

Lembke, Anna. 2023. *Dopamine Nation: Finding Balance in the Age of Indulgence.* New York: Dutton.

Lemoine, Blake. 2023. "'I Worked on Google's AI. My Fears Are Coming True.'" *Newsweek* (February 27). www.newsweek.com/google-ai-blake -lemoine-bing-chatbot-sentient-1783340.

Let Grow. n.d. "Playing Outside Should Not Be against the Law." Let Grow. Accessed January 14, 2024. https://letgrow.org/program/policy -and-legislation/.

Liberto, Daniel. 2023. "Anti-Fragility: Definition, Overview, FAQ." Investopedia. Last updated October 12, 2023. www.investopedia.com/terms/a /anti-fragility.asp.

Listrak. n.d. "Predictive Personalization Strategies." Listrak. Accessed August 18, 2023. www .listrak.com/white-papers/predictive -personalization-strategies.

Mac, Ryan, Sheera Frenkel, and Kevin Roose. 2022. "Skepticism, Confusion, Frustration: Inside Mark Zuckerberg's Metaverse Struggles." *New York Times* (October 9). www .nytimes.com/2022/10/09/technology /meta-zuckerberg-metaverse.html.

References   299

Magliano, Joseph. 2015. "Why Are Teen Brains Designed for Risk-Taking?" *Psychology Today* (June 9). www.psychologytoday.com/us/blog /the-wide-wide-world-psychology /201506/why-are-teen-brains-designed -risk-taking.

Meta. 2022. "Terms of Use." Instagram Help Center. https://help.instagram .com/581066165581870.

Meta. 2023a. "Instagram Quiet Mode: A New Way to Manage Your Time and Focus." Meta. https://about.fb.com /news/2023/01/instagram-quiet-mode -manage-your-time-and-focus/.

Meta. 2023b. "Reducing the Distribution of Problematic Content." Meta Transparency Center. https:// transparency.fb.com/enforcement /taking-action/lowering-distribution -of-problematic-content/.

Meta. n.d.a. "At Meta, We're Committed to Giving People a Voice and Keeping Them Safe." Meta Transparency Center. Accessed August 18, 2023. https:// transparency.fb.com/.

Meta. n.d.b. "Bullying and Harassment." Meta Transparency Center. Accessed August 18, 2023. https://transparency.fb.com/policies /community-standards/bullying -harassment.

Meta. n.d.c. "Misinformation." Meta. Accessed January 16, 2024. https://transparency.fb.com /policies/community-standards /misinformation#policy-details.

Mohan, Neal. 2021. "Perspective: Tackling Misinformation on YouTube."

YouTube Official Blog (August 25). https://blog.youtube/inside-youtube /tackling-misinfo/.

Moller-Nielson, Thomas. 2022. "What Are Our Phones Doing to Us?" *Current Affairs* (August 14). www.currentaffairs .org/2022/08/what-are-our-phones -doing-to-us.

Monteverde, Danny. 2020. "Fourth Grader's Suspension for BB Gun Seen on Zoom Call Reduced in Hearing after Namesake Legislation." WWL-TV (December 4). www.wwltv .com/article/news/crime/bb-gun -suspension-zoom-call-reduced/289 -3ebd6811-2b38-426d-8219-13a8fcf2c2b9.

Morrison, Ryan. 2021. "Step Away from Your Phone! Staring at Screens for Too Long 'Increases Risk of Short-Sightedness by Up to 80%'—and Half the World May Need Glasses by 2050, Study Warns." *Daily Mail* (October 7). www.dailymail.co.uk/sciencetech /article-10070219/Staring-phone -screens-long-increase-risk-short -sightedness.html.

Mühlberg, Byron. 2020. "Two Children Take on Google over Biometric Data." *CPO Magazine* (April 15). www.cpomagazine.com /data-privacy/two-children-take-on -google-over-biometric-data/.

Murphy, Mike. 2019. "This App Is Trying to Replicate You." *Quartz* (August 29). https://qz.com/1698337/replika-this-app -is-trying-to-replicate-you.

Murphy, Samantha. 2022. "TikTok May Show Harmful Content to Teens in Minutes." WRAL TechWire. https://wraltechwire.com/2022/12/16

/tiktok-may-push-potentially-harmful
-content-to-teens-within-minutes-study
-finds/.

Musiart. 2021. "Think of Someone, If
Your Pupil Expands It's a True Love."
YouTube (March 20). www.youtube
.com/watch?v=AreVHoJU75U.

National Institute of Mental Health.
2023. "Mental Illness." National
Institute of Mental Health. Last
updated March 2023. www.nimh
.nih.gov/health/statistics/mental
-illness#part_155771.

Newman, Susan. 2017. "Study
Underscores Why Fewer Toys Is
the Better Option." *Psychology Today*
(Decembrer 14). www.psychology
today.com/us/blog/singletons/201712
/study-underscores-why-fewer-toys-is
-the-better-option.

Newport, Cal. 2022. "On Teenage
Luddites." Calnewport.com
(December 16). https://calnewport
.com/on-teenage-luddites/.

**Norton, Michael I., Daniel Mochon,
and Dan Ariely.** 2012. "The IKEA
Effect: When Labor Leads to Love."
Journal of Consumer Psychology 22,
no. 3 (July). www.hbs.edu/ris
/Publication%20Files/norton%20
mochon%20ariely_6f7b1134-06ef-4940
-a2a5-ba1b3be7e47e.pdf.

**National Wildlife
Federation.** n.d. "Certify Your Habitat
to Help Wildlife!" National Wildlife
Federation. Accessed August 18, 2023.
www.nwf.org/certifiedwildlifehabitat.

O'Connor, J. J., and E. F. Robertson.
1999. "Abu Ja'far Muhammad ibn

Musa Al-Khwarizmi," MacTutor (July),
https://mathshistory.st-andrews.ac.uk
/Biographies/Al-Khwarizmi/.

O'Connor, Lydia. 2013. "San Francisco
Train Passengers Too Distracted by
Phones to Notice Shooter's Gun in
Plain Sight." *HuffPost* (October 8).
www.huffpost.com/entry/san
-francisco-train-shooting_n_4066930.

Odell, Jenny. 2019. *How to Do Nothing:
Resisting the Attention Economy.*
Brooklyn, NY: Melville House.

**O'Keeffe, Gwenn Schurgin, and
Kathleen Clarke-Pearson.** 2011.
"The Impact of Social Media on
Children, Adolescents, and Families."
Pediatrics 127, no. 4. https://publications
.aap.org/pediatrics/article/127/4/800
/65133/The-Impact-of-Social-Media-on
-Children-Adolescents.

Okunytė, Paulina. 2023. "Instagram
Influencers with up to 40m Followers
Use Russian Bots." Cybernews
(November 15). https://cybernews
.com/security/instagram-influencers
-use-russian-bots-data-leak-shows/.

Ortiz, Camilo. 2023. "Treating
Childhood Anxiety with a Mega-Dose
of Independence." *Profectus* (March 14).
https://profectusmag.com/treating
-childhood-anxiety-with-a-mega-dose
-of-independence/.

Outside Play. n.d. "Frequently Asked
Questions." OutsidePlay.ca. Accessed
August 18, 2023. https://outsideplay
.ca/#/faq.

Park, Elaine. 2018. "Q&A: Anna
Lembke on Smartphone Technology
Addiction." *Stanford Daily*

(February 22). https://stanforddaily
.com/2018/02/22/qa-anna-lembke-on
-smartphone-technology-addiction/.

PBS. 2018. "The Facebook Dilemma,
Part 1." *Frontline.* www.pbs.org/wgbh
/frontline/documentary/facebook
-dilemma/.

Pelley, Scott. 2023. "Is Artificial
Intelligence Advancing Too Quickly?
What AI Leaders at Google Say." *60
Minutes.* www.cbsnews.com/news
/google-artificial-intelligence-future
-60-minutes-transcript-2023-04-16/.

Phillips Exeter Academy. n.d.
"Harkness." Phillips Exeter Academy.
Accessed August 18, 2023. www.exeter
.edu/excellence/how-youll-learn.

Pilkington, Caitrin. 2022. "How
TikTok's Pupil Test Works, and What
It Tells Us about Love and Attraction."
Pulse (February 11). https://whyy.org
/segments/how-tiktoks-pupil-test
-works-and-what-it-tells-us-about-love
-and-attraction/.

Price, Catherine. 2018. "Trapped—the
Secret Ways Social Media Is Built to
Be Addictive (and What You Can Do
to Fight Back)." BBC Science Focus
(October 29). www.sciencefocus
.com/future-technology/trapped-the
-secret-ways-social-media-is-built-to
-be-addictive-and-what-you-can-do-to
-fight-back/.

Prothero, Arianna. 2023. "Kids' Screen
Time Rose during the Pandemic
and Stayed High. That's a Problem."
Education Week (February 28). www
.edweek.org/leadership/kids-screen
-time-rose-during-the-pandemic-and
-stayed-high-thats-a-problem/2023/02.

**Rajanala, Susruthi, Mayra B. C.
Maymone, and Neelam A. Vashi.**
2022. "Selfies—Living in the Era of
Filtered Photographs." *JAMA Facial
Plastic Surgery* 20, no. 6 (November 15).
doi.org/10.1001/jamafacial.2018.0486.

Redick, Scott. 2013. "Surprise Is Still
the Most Powerful Marketing Tool."
Harvard Business Review (May 10).
https://hbr.org/2013/05/surprise-is
-still-the-most-powerful.

Reilly, Kaitlin. 2023. "Why Some
Parents Are Committing to Spending
1000 Hours Outside This Year: 'My
Kids Fall Asleep a Lot Faster.'" Yahoo
(March 17). www.yahoo.com/lifestyle
/1000-hours-outside-challenge
-151909967.html.

Rosales, Isabel. 2019. "Sex Offender
Arrested after Using the Bible App to
Target Teen Girls at a Hillsborough
County Church." *ABC Action
News* (October 31). https://www
.abcactionnews.com/news/region
-hillsborough/sex-offender-arrested
-after-using-the-bible-app-to-target
-teen-girls-at-a-hillsborough-county
-church.

Rushkoff, Douglas. 2011. *Program or Be
Programmed: Ten Commands for a Digital
Age.* New York: Soft Skull.

Rushkoff, Douglas. 2014. *Present Shock:
When Everything Happens Now.* New
York: Penguin.

Sachs, Sam. 2022. "Self-Harm
Mentions up 500% on Twitter;
Which Hashtags to Watch." WFLA
(August 31). www.wfla.com/news
/national/self-harm-mentions-up-500
-on-twitter-which-hashtags-to-watch/.

Sadowski, Jathan. 2021. "Facebook Is a Harmful Presence in Our Lives. It's Not Too Late to Pull the Plug on It." *Guardian* (October 6). www.the guardian.com/commentisfree/2021/oct/06/facebook-scandals-social-media.

SafeAtLast. 2022. "Intriguing Amazon Alexa Statistics You Need to Know in 2023." SafeAtLast. https://safeatlast.co/blog/amazon-alexa-statistics/#gref.

Sager, Jeanne. n.d. "These Are the TikTok Challenges Every Teacher Should Know About." TeachStarter. Last updated April 2022. www.teachstarter.com/us/blog/school-tiktok-challenges-kids-doing/.

Salganik, Matthew J., Peter Sheridan Dodds, and Duncan J. Watts. 2006. "Experimental Study of Inequality and Unpredictability in an Artificial Cultural Market." *Science* 311 (5762): 854–56. https://doi.org/10.1126/science.1121066.

Sax, David. 2016. *The Revenge of Analog: Real Things and Why They Matter.* New York: PublicAffairs.

Schmoon. n.d. "Twenty-three Phrases to Sound Like You Live in Silicon Valley." Superside. Accessed August 18, 2023. www.superside.com/blog/silicon-valley-phrases.

Sellers, Dennis. 2002. "Apple, iBooks, PowerSchool Make School News." Macworld (October 10). www.macworld.com/article/156283/school-2.html.

Shellenbarger, Sue. 2016. "Most Students Don't Know When News Is Fake, Stanford Study Finds." *Wall Street Journal* (November 21). www.wsj.com/articles/most-students-dont-know-when-news-is-fake-stanford-study-finds-1479752576.

Shifrin, Donald, Ari Brown, David Hill, Laura Jana, and Susan K. Flynn. 2015. "Growing Up Digital: Media Research Symposium." American Academy of Pediatrics (October 1). https://studylib.net/doc/12041195/growing-up-digital.

Singer, Emma. 2023. "When Should Kids Get TikTok? Here's What Experts Say." PureWow (February 16). www.purewow.com/family/when-should-kids-get-tiktok.

Skenazy, Lenore. 2020a. "Cops Arrest Brooklyn Rabbi for Letting Kids— Ages 11, 8, and 2—Walk to the Store." *Reason* (May 12). https://reason.com/2020/05/12/cops-arrest-brooklyn-rabbi-for-letting-kids-ages-11-8-and-2-walk-to-the-store/.

Skenazy, Lenore. 2020b. "Public School Forces Kids to Take the Bus Home. Walking Is Faster." *Reason* (November 10). https://reason.com/2020/11/10/spann-elementary-walking-home-school-kids.

Smith, Katie Louise. 2021. "TikTok's Side Eye Challenge: The Viral Trend Explained." PopBuzz (March 10). www.popbuzz.com/internet/viral/tiktok-side-eye-challenge-explained/.

Smithsonian Migratory Bird Center. n.d. "Bird Memory Phrases." Smithsonian National Zoological Park. Accessed January 30, 2024. https://nationalzoo.si.edu/sites/default/files/documents/bird_memory_phrases.pdf.

Social Media TestDrive. n.d. "Education for the Digital Age." Social Media TestDrive. Accessed August 18, 2023. https://socialmediatestdrive.org/.

Solnit, Rebecca. 2001. *Wanderlust: A History of Walking.* New York: Penguin.

Solnit, Rebecca. 2014. "Poison Apples." *Harper's* (December). https://harpers.org/archive/2014/12/poison-apples/.

Sommerfeld, Julia. n.d. "Human Brain Gets a Kick out of Surprises." CCNL. Accessed August 18, 2023. www.ccnl.emory.edu/Publicity/MSNBC.HTM.

Steinbrenner, Corinne. 2022. "Why Do Some Kids Take Bigger Risks Than Others?" *The Brink* (September 29). www.bu.edu/articles/2022/why-do-some-kids-take-bigger-risks-than-others/.

Steiner-Adair, Catherine, with Teresa H. Barker. 2014. *The Big Disconnect: Protecting Childhood and Family Relationships in the Digital Age.* New York: HarperCollins.

Stephens, M., et al. 2015. "Is Reading Contagious? Examining Parents' and Children's Reading Attitudes and Behaviors." *Policy Brief* no. 9 (December). www.researchgate.net/publication/299346391_Is_reading_contagious_Examining_parents'_and_children's_reading_attitudes_and_behaviors.

Storm, Julia. 2021. "Parenting in the Digital Age: How to Teach Our Kids Healthy Tech Habits." Freedom (June 14). https://freedom.to/blog/healthy-tech-habits-kids.

Strauss, Neil. 2016. "Why We're Living in the Age of Fear." *Rolling Stone* (October 6). www.rollingstone.com/politics/politics-features/why-were-living-in-the-age-of-fear-190818/.

Szathmary, Zoe. 2017. "Having It His Way! Eight Old Boy, Craving a Cheeseburger Drives Himself and His Four-Year-Old Sister to McDonald's while Their Parents Sleep." *Daily Mail.* www.dailymail.co.uk/news/article-4406544/Eight-old-boy-craving-cheeseburger-drives-McDonald-s.html.

Taleb, Nassim N. 2014. *Antifragile: Things That Gain from Disorder.* New York: Random House.

Thorn. 2022. "Responding to Online Threats: Minors' Perspectives on Disclosing, Reporting, and Blocking." www.thorn.org/thorn-research-minors-perspectives-on-disclosing-reporting-and-blocking/.

TikTok. 2022. "Introducing More Ways to Create and Connect with TikTok Now." TikTok (September 15). https://newsroom.tiktok.com/en-us/introducing-tiktok-now.

TikTok. 2023. "Death Row Is in the (TikTok) House!" TikTok (February 14). https://newsroom.tiktok.com/en-us/death-row-is-in-the-tiktok-house.

TikTok. n.d. "Looking for Trends? You've Come to the Right Place." Accessed January 16, 2024. https://ads.tiktok.com/business/creativecenter/trends/home/pc/en.

TikTok Fun. 2022. "Happiness Latest Is Helping Good Kids TikTok Videos 2022 | A Beautiful Moment in Life #9." YouTube. www.youtube.com /watch?v=R_61NMIVk4s.

Toole, Brittany. n.d. "Risky Play for Children: Why We Should Let Kids Go Outside and Then Get out of the Way." CBC. Accessed August 18, 2023. www .cbc.ca/natureofthings/features/risky -play-for-children-why-we-should-let -kids-go-outside-and-then-get-out.

Trapani, Gina. 2007. "'Don't Break the Chain' to Build a New Habit." Lifehacker (June 24). https://lifehacker .com/jerry-seinfelds-productivity -secret-281626.

Treisman, Rachel. 2022. "The FBI Alleges TikTok Poses National Security Concerns." NPR (November 17). www .npr.org/2022/11/17/1137155540/fbi -tiktok-national-security-concerns-china.

Trilling, David. 2017. "Smartphones Are Distracting Even When You're Not Using Them." *Journalist's Resource* (July 27). https://journalistsresource .org/economics/smartphones-mobile -distracting-cognition-iphone/.

Turkle, Sherry. 2016. *Reclaiming Conversation: The Power of Talk in a Digital Age.* New York: Penguin.

Turner, Cory. 2022. "Six Things We've Learned about How the Pandemic Disrupted Learning." *All Things Considered* (June 22). www.npr.org /2022/06/22/1105970186/pandemic -learning-loss-findings.

Twenge, Jean M., Thomas E. Joiner, and Gabrielle N. Martin. 2017.

"Increases in Depressive Symptoms, Suicide-Related Outcomes, and Suicide Rates Among U.S. Adolescents after 2010 and Links to Increased New Media Screen Time." *Association for Psychological Science* 6, no. 1. https:// journals.sagepub.com/doi/10.1177 /2167702617723376.

Vadukul, Alex. 2022. "'Luddite' Teens Don't Want Your Likes." *New York Times* (December 15). www.nytimes .com/2022/12/15/style/teens-social -media.html.

Van Geel, Mitch, Paul Vedder, and Jenny Tanilon. 2014. "Relationship between Peer Victimization, Cyberbullying, and Suicide in Children and Adolescents: A Meta-Analysis." *JAMA Pediatrics.* https://jamanetwork .com/journals/jamapediatrics /fullarticle/1840250.

Vincent, James. 2021. "Tom Cruise Deepfake Creator Says Public Shouldn't Be Worried about 'One-Click Fakes.'" *Verge* (March 5). www .theverge.com/2021/3/5/22314980 /tom-cruise-deepfake-tiktok-videos-ai -impersonator-chris-ume-miles-fisher.

Waldo Photos. n.d. "Terms of Use and Privacy Policy." Waldo Photos. Accessed August 18, 2023. https:// waldophotos.com/terms-of-use/?.

Walker, Tim. 2023. "Cellphone Bans in School Are Back. How Far Will They Go?" National Education Association. www.nea.org/advocating-for-change /new-from-nea/cellphone-bans-school -are-back-how-far-will-they-go.

Wallace, Kelly. 2015. "Forget Sitting Down! Students Now Standing Up to

Learn." CNN (December 10). www
.cnn.com/2015/12/10/health/standing
-desks-impact-health-education/index
.html.

Walsh, Erin. 2023. "Mental Health,
TikTok, and Teens." Spark and
Stitch Institute (January 19). https://
sparkandstitchinstitute.com/mental
-health-tiktok/.

Weiss, Dawid. 2021. "Historic Slogans
and Claims of IT and Computer
Brands." Neuroflash (August 31).
https://neuroflash.com/blog/slogans
-claims-of-it-computer-brands/.

Wikipedia. n.d.a. "Alan Turing."
Wikipedia. Accessed August 18, 2023.
https://en.wikipedia.org/wiki/Alan
_Turing.

Wikipedia. n.d.b. "Argument from
Authority." Wikipedia. Accessed
August 18, 2023. https://en.wikipedia
.org/wiki/Argument_from_authority.

Wikipedia. n.d.c. "Marvin Minsky."
Wikipedia. Accessed August 18, 2023.
https://en.wikipedia.org/wiki/Marvin
_Minsky.

WKBW. 2022. "I Wish My Mom's
Cell Phone Was Never Invented:
Child's Essay Goes Viral." WTMJ-TV
Milwaukee. www.tmj4.com/news
/national/i-wish-my-mom-s-cell-phone
-was-never-invented-child-s-essay-goes
-viral.

Wong, Brittany. 2021. "Wait,
What the Heck Is a 'Parasocial

Relationship'?" *HuffPost* (May 26).
www.huffpost.com/entry/parasocial
-relationships-with-celebrities_l
_60a56a18e4b0d45b75248115.

**Wouters, Niels, and Jeannie M.
Paterson.** 2021. "TikTok Captures
Your Face." Pursuit (July 26). https://
pursuit.unimelb.edu.au/articles/tiktok
-captures-your-face.

WSJ. 2021. "Teen Mental Health
Deep Dive." *Wall Street Journal*
(September 29). https://s.wsj.net/public
/resources/documents/teen-mental
-health-deep-dive.pdf.

Xanadu87. 2022. "How to Disable
'One More Minute' Screen Time
Limit on iPhone?" Apple Discussions
(February 12). https://discussions.apple
.com/thread/253661766.

Xiang, Chloe. 2023. "ChatGPT Is
Passing the Tests Required for Medical
Licenses and Business Degrees." VICE
(January 23). www.vice.com/en/article
/akebwe/chatgpt-is-passing-the-tests
-required-for-medical-licenses-and
-business-degrees.

Yates, Eames. 2017. "What Happens
to Your Brain When You Get a Like
on Instagram." *Business Insider*
(March 25). www.businessinsider
.com/what-happens-to-your-brain-like
-instagram-dopamine-2017-3.

Zuckerberg, Mark. 2021. "Founder's
Letter, 2021." Meta (October 28).
https://about.fb.com/news/2021/10
/founders-letter/.